WORKING WITH WATER

水之匠心

（美）斯科特·休斯　（美）约翰·乌姆班霍瓦尔／著
张晨／译

休斯乌姆班霍瓦尔建筑事务所作品集
hughesumbanhowar Architects

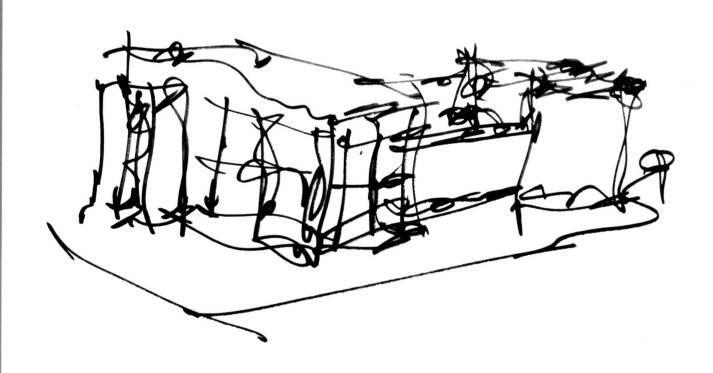

辽宁科学技术出版社
·沈阳·

hughesumbanhowar Architects

TABLE OF CONTENTS
目录

Preface / 前言   004
    by Pierluigi Serraino   皮埃路易吉·赛拉诺

Beach Road 2 / 海滨路 2 号   008

Big Timber / 大廷伯河畔休闲住宅   018

Stone / 斯通宅邸   038

East Meets West / 东方与西方相遇   050

Paseo Miramar / 帕萨奥·米拉玛   062

Baker's Bay / 贝克湾   070

Grenville / 格伦维尔   080

Pine School / 潘恩学校   094

Tasting Room / 品酒屋   106

River Road / 河畔别墅   116

One for One / One for One 住宅   132

Hutchinson Island / 哈钦森岛   144

Ski House / 滑水屋   152

Holly Lane / 冬青道   164

Strulowitz / Strulowitz 住宅   182

Nests / 筑巢   192

Oblivious Office Building / 隐秘的办公大楼   202

Abbott Kinney / 阿尔伯特·金尼别墅   208

Prairie Chapel / 草原小教堂   218

Augusta / 奥古斯塔   226

Erik / 埃里克   240

PREFACE
前言

ARCHITECTURE OF MODERN CLIMATES.
Notes on the work of HUUM Architects.
Pierluigi Serraino

California and Florida are two major theaters in the meteoric ascent of Modern Architecture in the United States. To set up shop in both regions is a symbolic commitment to growing a practice rooted in that pioneering modernity, and to update its currency to the priorities of our time. Such an uncommon bicoastal arrangement brings the prominence of the indoor-outdoor design theme to the forefront with this duo, an ongoing test for architects of any generation to engage the natural and built world in ever- changing ways.

Scott Hughes and John Umbanhowar, founders of HUUM architects, have scripted slices of the modern legacy into their very own histories. Scott, an East Coast native, grew up breathing the powerful monumentality of master architect Louis Kahn. He regularly visited his uncle, renowned scholar in History of Technology Thomas Hughes, at his residence in the highly singular and celebrated Vanna Venturi House by Robert Venturi. Stints in the offices of Philip Johnson and Hugh Newell Jacobsen made him aware that Modern Architecture, rather than being a universal formula with a single response to all site and human conditions, offered as many permutations as the larger-than-life personalities heading their own offices in localities whose architecture was coming into being. John, raised in Minnesota, arrived in California in the mid-90's as an Art History major with a penchant for challenging the authority of the canon. For him, in the monolithic narrative of aesthetic hierarchies and the lineage of the masters, anything and anyone could be questioned—and it invariably would. His first encounter with architecture took place before his undergraduate days, when during a cold winter he ventured into the St. Paul Abbey in Collegeville, Minnesota. This Benedictine monastery carrying the signature of venerable Bauhaus architect Marcel Breuer allowed him to experience the sculptural assertiveness of architectural form firsthand, and emblazoned the memory of that lesson in his design ventures.

These two parallel histories found their common ground at the Southern California Institute of Architecture (SCI-Arc) in Los Angeles. Graduate school represented for one a re-engagement with the larger ambition of architecture as an art form, while for the other it offered the opportunity to bridge the knowledge of architecture as a finished artifact to the process of generating it. SCI-Arc in the mid-90's was steeped in object-making and the studio was the locus where much of that exploration took place. That free play of material handling, their connections, and the discovery of the expressive potential inherent in the physicality of things was paramount; the institution was instrumental in developing tectonic consciousness in its students before its leadership uncompromisingly embraced an all-encompassing digital turn at the dawn of the new millennium. It is the cursory account of these separate

现代风潮中的建筑
HUUM 建筑事务所作品札记
皮埃路易吉 · 赛拉诺

加利福尼亚州和佛罗里达州是美国现代风格建筑迅速发展的两个主要地点。能在这两地建立办事处象征着对以先进现代性为基础的这个领域的全情投入，也是根据这个时代的优先顺序进行的更新与修正。将办事处分设在美国东西海岸这样不同寻常的安排使得室内外设计主题的优势凸显。如何以不断变化的形式适应自然环境与建成环境，是对所有世代建筑师的持续考验。

HUUM 建筑事务所的创始人斯科特 · 休斯与约翰 · 乌姆班霍瓦尔已经将现代建筑遗产的点点滴滴写入了自己的历史。斯科特在美国东海岸出生，呼吸着伟大建筑师路易斯 · 卡恩强而有力的标志性建筑气息长大。他经常到叔叔——科学技术史著名学者托马斯 · 休斯的家中做客。那是由罗伯特 · 文丘里设计的非常特别且十分著名的"母亲之家"。菲利普 · 约翰逊和休 · 纽尔 · 雅各布森办公室的节约感让他在筹备自己事务所标榜传奇个性的办公室时，意识到现代建筑风格提供了尽可能多的组合形式，而不是用统一模式解决全球范围内所有场地情况与人力情况下的问题。约翰在明尼苏达州长大，20 世纪 90 年代中期来到加州进行艺术史专业的学习，喜欢挑战权威。对他来说，在审美层次的整体叙述以及大师的世界里，任何事、任何人都可以被质疑，也总是会受到质疑。他与建筑的第一次接触发生在上大学之前：在一个寒冷的冬天，约翰偶然来到了明尼苏达州科利奇维尔的圣保罗大教堂。这间本笃修道院具有包豪斯建筑师马塞尔 · 布鲁尔标志性特征，而它使得约翰亲身体验到建筑形式的雕塑感魅力，并在他的设计实践中对这段回忆加以装饰升华。

这两段平行的经历在洛杉矶的南加州建筑学院发生了交集。研究生院代表的一方面是与更具野心的建筑艺术形式的关系重启，另一方面提供了将建筑知识的成果与创作过程连接的机会。90 年代中期，南加州建筑学院专注于各种项目，而工作室正是大部分探索活动发生的地点。材料处理、材料连接以及发掘事物内在表现潜力中的自由发挥占据了重要的地位；在新千年开始之际坚决地接受全面的数字化转型之前，南加州建筑学院帮助发展学生的建筑意识。HUUM 建筑事务所作为全美众多年轻的建筑事务所中不断成长的一个，创始人两人各自的发展故事以及在南加州建筑学院相遇的粗略讲述提供了解读并理解事务所作品的关键。

斯科特与约翰的世界充满坐标轴和深远的视线。大胆的几何图形与悬臂造型是同时具有辨识性和独特性的设计语言。在他们

origins, and of their point of contact in Southern California, that offers the key to read and understand the architectural production of HUUM, an office with a growing presence in the constellation of young practices on the national scene.

Theirs is a world of axes and deep sightlines. Bold geometries and cantilevers are some of the words in a design vocabulary both recognizable and distinctive. In their projects, frames and decks reappear as signature quotations but tell a different spatial story every time. Double heights and wide openings find fluid implementation in many schemes. Even more importantly, theirs is in architecture of deep materiality, awakening all human senses. The surfaces and textures of water bring about bodily awareness in those experiencing the space HUUM has designed around them. It is a universe filled with phenomenological stimuli, that is, spatial conditions eliciting a reaction from all human senses, and the cultivated echoing of mythological precedents. While the references are traceable, the work is filled with contemporary aspirations. It is a relaxed formula of total architecture, where leisure and control inhabit each proposition in equal amounts. Both rigorous and intelligible, exact and amicable, their spaces choreograph public and private realms without intimidating occupants with the invasive presence of a design idiom hijacking collective attention at the expense of a building's intended day-to-day functionality. The absence of an inflexible prescription, the type of one-solution for all cases so dear to the early modernists and so condemned later in the 20th century, allows for a fruitful dialog with the vernacular when the intersection of patron and site-specificity calls for that kind of molecular attention.

For architects committed to their art, projects are like children. They all carry the genes of their creators even when maintaining marked differences in physical appearance and temperamental traits. In this vein, the Miesian rigor of the Hutchinson Island house, a quasi-constructivist fantasy of overlapping bars placed in tropical coordinates, coexists coherently with the Gaudiesque pitch of the curvilinear footprint defining the dominant image of the Grenville residence. If in the former, the horizontal and vertical cleavage between the interlocking volumes reveals Nature beyond, while in the latter the topography of roofs and curvilinear perimeter determines its own logical hardscape to both encompass and complement its lush surroundings. While both houses feature shutters for sun shading, their impact on the overall result is markedly dissimilar.

In HUUM's hands, modernist motifs are consistently revisited and reinterpreted. The frame and the box, archetypical signs—when pure—shared within the generation of Walter Gropius and his followers (especially in the Northeast) are intervened upon. Their outlines have creases, like in the lifted volume crowing the bar of Big Timber Riverside in Montana, and chamfers, like in Beach Road 2 and The Nest. That expressive plasticity is the hallmark of a healthy relationship between the present and its origins; a tribute to what was with overtures to what

的项目中，构架和平台是标志性设计元素，但每次重复出现讲述的都是不同的空间故事。高度与宽度加倍的窗口在许多项目中得到了优雅流畅的表现。更重要的是，他们的作品具备深度的物质性，能够唤醒所有的人体感官。水景的外观及纹理让人们在 HUUM 建筑事务所设计的空间中感受到身体意识的觉醒。这是一个充满现象性刺激元素的空间，具备诱发所有人体感官反应的空间条件以及虚构先例的文雅启示。尽管设计中的参考对象可查，但设计仍充斥着现代的渴望与追求。它是形式松散的总体结构，每个方案中放松与控制的考量旗鼓相当。既严密又易懂，既准确又友善，他们的公共及私人空间设计安排并不会因为以牺牲建筑的日常功能为代价追求公众注意力的设计习惯让居住者感到敬而远之。早期现代主义者珍视的以不变应万变而缺乏灵活性的解决方案在 20 世纪后期备受责难。但这种设计形式使得赞助者与区位特性产生的交集需要某种关注时，设计能够与当地环境之间进行成果丰硕的对话与交流。

对于那些热爱工作的建筑师来说，项目就像自己的孩子一样。即便在外观和风格上保留着明显区别，它们也同样承载着创造者的基因。在这个层面上，哈钦森岛住宅的类构成主义密斯式严谨设计得以与作为格伦维尔住宅主要特征的曲线造型共存。如果前者交错体量间的水平与垂直缝隙展现出远处的自然景色，后者中的屋顶形态与曲线造型则决定了自身硬景观的合理性，既包含郁郁葱葱的环境又对其进行了补充。两间住宅都配置了遮阳百叶窗，呈现的整体效果却截然不同。

经 HUUM 建筑事务所之手，现代主义不断获得重新审视和重新解读。框架和盒子等被沃尔特·格罗皮乌斯和他的追随者（尤其是在东北地区）世代共有的典型标识被加工处理。他们的外形具有褶皱，就像在蒙大拿的大廷伯河畔休闲住宅以及海滨路 2 号和"筑巢"等项目中的倒角设计那样。极富表现力的可塑性特质是现在与起源之间健康关系的标志；一次向着未来的致敬。这种对话在理查德·纽佐尔的帕萨奥·米拉玛重建项目中变得更为明确。这个项目作为现代主义经典设计充分展现了这位奥地利建筑先锋的设计语言。处理这处加州现代主义地标结构时，模块化的设计带来充足的建筑可能性，对设计师来说也是一个惊喜。HUUM 建筑事务所的项目并非奉行严格保护策略，而是作为一种干预手段满足新主人的各种需求。在查看纽佐尔的项目原始图纸时，斯科特和约翰领悟了纽佐尔经典设计的真谛；同时，他们也提炼出自己信仰，认为建筑既可以与人类交流，也可以宣扬其居民的独特性。在这种情况下，光束扩展与浮动屋顶两种标志性加州元素，都作为建筑中的背景元素朝向自然开放。

尽管作品集中收录的主要为独栋房屋，HUUM 建筑事务所的审美

will be. That dialog becomes even more explicit in the restoration of Richard Neutra's Paseo Miramar, a modernist classic showcasing the full design vocabulary of the Austrian pioneer. That modularity could be rich with architectural possibilities was a surprise even to them as they surgically handled the body of a landmark piece of California Modernism. Rather than an exercise in strict preservation, theirs was an intervention as philologically accurate as it was current in addressing the needs of a new owner. In examining Neutra's original drawings, Scott and John understood the DNA of his timeless visions; in turn, they refined their own beliefs on how architecture can both interface with human existence and celebrate the uniqueness of its inhabitants. Beam extensions and floating roofs, both trademarks of the California look, are in this instance background components of an architecture open to Nature.

While their portfolio largely features single-family homes, their aesthetic finds cogent outlet in other building types: a school, an helicopter hangar, a deconstructed quonset hut turned chapel in the prairie, a wine-tasting room in Italy, and an office building in American suburbia. These are instances of a formal world amenable to encompass functions and briefs highly diverse from each other, across the country and abroad. It is the inherent malleability of their design language that lends itself to credible renditions in places ordinarily unlikely to be available in an architect's career. Planes can arch upward and bend downward, yet still belong to a family of shape understandable in their design message. Ultimately, producing constantly varying architectural results while remaining recognizable in each iteration is where a design architect exhibits command in space-making. Grids, cubic massing, sticks, fanned geometries, folded planes and the stacking of volumes are some of the devices adopted to make bold statements in the landscape. As the buildings extend into the outdoor, the design language remains the same and what was enclosed becomes hardscape as it meets the elements. It is a subtle gradation of changes of levels, cascading platforms and parades of ceiling heights, always logical and always mastered in architectural implications.

Experiencing these designs, it is immediately apparent to the viewer that a house conceived for the state of Florida carries marked differences from one built in California, even though they both face their respective coastlines in Southern climates. It is the management of environmental conditions, how the wind is broken within the crevices of the house's exterior areas and how the humidity versus the dryness of the air affects the openings for passive ventilation, which reveals the project's location. That authenticity is a constant of the modernist ethos, but in the case of Scott and John it is heartfelt rather than being the dogmatic following of a precept; that loose relationship is what brings lyricism into much of their work. It is about living in a house as opposed to being enclosed in architecture as a work of art impervious to the needs of its occupants. The traceable transparency within all their schemes is synonymous with the porosity of architecture as it touches the ground. Boundaries are evoked through glass instead of solid walls, which reject the outside world. That friendliness is what makes them welcoming, like pavilions on the beach, bestowing a diffused luminosity in all quarters of their projects.

准则在其他建筑类型中找到了极具说服力的表现形式：一所学校、一间直升机库、一间解构拱屋改造而成的草原小教堂，意大利的一间品酒屋，以及美国郊区的一栋办公建筑。这些都是分布在美国与世界各地，包含高度多样化的功能与概要的案例。设计语言的内在适应性引导其自身在往常不太可能实现建筑工程的地方发挥可靠表现。在HUUM建筑事务所的设计中，楼面可以向上拱起或向下弯曲，但在他们的设计语言中仍属于可以理解的形状范畴。最终，形成不断变化的建筑设计方案，同时保证每个迭代版本中原型可辨，是建筑师在空间设计中对项目控制的体现。可以利用网格、立方体块、棍棒、扇形几何体、折叠平面和体量堆叠等方法创作大胆的景观设计。随着建筑延伸到室外，设计语言保持一致，原本封闭的结构与元素结合成为了硬景观。这是一个结构高度的微妙渐变过程，拾级而下的平台与变化的天花板高度总是保持逻辑感和建筑寓意。

体验了这些设计以后，观察者可以立刻清楚地发现建在佛罗里达州的房屋与建在加利福尼亚州的住宅即使都处在南部气候之中，分布朝向海边，也存在着明显的设计差异。风是如何在房屋外部的缝隙处分散开，空气的湿度与干燥度如何影响被动通风的通风口设计——这些环境条件的管理揭示了项目的地理位置差异。真实性是现代主义思潮中的一个常数，但在斯科特与约翰的身上，这是发自内心的感悟而不是对某条格言的教条遵从；这种松散的关系为他们的大部分作品注入了情怀。毕竟生活发生的地点是家，而不是封闭的建筑，不是对居住者的需求无动于衷的艺术品。在他们的所有项目中可循的通透性在实施之后便是建筑孔隙率的代名词。玻璃代替厚实的墙壁划出空间界线，避免了对外界的隔离。项目展现出的友好使它们备受欢迎，如沙滩上的凉亭，各个角落都充满了光亮。

无论在城市环境还是自然环境中，HUUM建筑事务所的设计都与所在环境构成动态互动。每个项目中的重大决策都放大了各个场地的潜力。阿尔伯特·金尼别墅是威尼斯海滩上一处教科书式的工程，展现出海滨路2号的联合建筑以及河畔别墅的双层高度建筑中存在的一致空间体验。项目本身很少，或者说几乎不以独栋的形式出现；相反它们是能够产生自身微城市性的复合式建筑，类似通过封闭通道连接的多层次结构。这样的项目方式追求计划层面的多样性以及同一主题的多种不同变化。这就是他们作品中的多样性与同一性。

截止到目前，HUUM建筑事务所的作品从不高于三层，贴近地面的特点非常明显。无论从基座还是较浅的底座建起，建筑结构都从平坦的地面上破土而出，抵消一楼平面以决定其独特的截面特征。即便是像维亚马里纳宅邸一样具有较小环境影响的项目，体量主导环境的形式也存在一定的纪念性。它的果敢自信是物质存在而非规模的副产品。"东方与西方相遇"项目则是现代古典主义的一个典型案例，四个结构围绕一个轴线，传递出比喻意义而非实际意义上的体量感。这些结构位于顶棚之下，

Whether in an urban district or in a natural setting, their designs interact dynamically with their contexts. Big moves displayed by each project magnify the potentiality of each site. The Abbot Kinney Loft is a textbook case of cultivated insertion in the vernacular of Venice Beach, yet exhibits the consistency of spatial experiences found in the link building of Beach Road 2 and in the double heights of the River Road House. The projects are themselves rarely, if ever, a single volume; instead they are compounds producing their own micro-urbanities, like citadels with a hierarchy of buildings connected through enclosed circulations. Such an approach yields regularities at a plan level and countless variations on a theme. This is where diversity and sameness inform their work.

Presently never taller than three stories, HUUM's architecture is decisively earthbound. Either rising from a podium or resting on a shallow plinth, the structures emerge from flat sites, offsetting that ground plane to determine their unique sectional distinction. Even when smaller in footprint, like in the Via Marina residence, there is a layer of monumentality in how the massing dominates its surroundings. Its assertiveness is the byproduct of its material presence as opposed to the result of scale. East Meets West is a paradigmatic example of modern classicism, with four pavilions grouped around an axis conveying a bigness which is metaphorical rather than factual. Volumes slide underneath canopies, leaving shadowlines intact to remove the weight from the physicality of architecture. Clerestory windows establish a datum that the eye follows with ease during the navigation of the space. It is a project of learned quotation from the vibrant legacy of mid-century modern while denying the viewer the ability to identify which precise antecedents are being quoted.

Confidence in the expressive import of Euclidean geometry informs to a large extent the firm's entire output. This hardly implies that alternative forms of space-making are unlikely to be entertained in this design group; it is a case-by-case consideration to be made when the situation presents itself. Either the site or the program or both would be primary drivers generative of a departure from an architecture growing out of orthogonal intersections, as much of their work reveals.

Architecture is a field in constant motion, and the exploration of new and alternative ways to make space is part and parcel of being an inquisitive architect. As the practice grows and commissions get bigger in scope and ambition, it is entirely possible to envisage this firm tackling additional themes such as the high-rise, the mixed-use development and the institutional building. Between Florida and California, two states continually resetting the bar of design excellence, HUUM architects is uniquely positioned to capitalize on this highly unusual bi-coastal arrangement, and the caliber of the clients this arrangement brings. The next five years will tell that story.

(Pierluigi Serraino is a practicing architect, author, and educator based in the San Francisco Bay Area.)

阴影的存在去除了建筑物质性中的重量感。天窗起到引导视线的空间作用。该项目是对20世纪中叶现代主义设计丰厚遗产的致敬，同时又让观看者无法辨别对应的具体内容。

对引入欧几里得几何学表达的信心影响了事务所项目的整体风格。这几乎就意味着不同形式的空间设计手法不可能在这个设计团队中被接受；情况出现时，要按具体条件进行分析判断。正如事务所的作品所呈现出的一样，场地、项目或二者同时都是发展中的建筑脱离主流的主要动力。

建筑是一个不断运动变化的领域。对空间设计新方法与替代方法的探索是一个勤奋建筑师不可或缺的品质。随着业务的累积，委托项目的规模和野心逐渐变大，HUUM建筑事务所已经完全具备处理高层建筑、多功能项目以及公共机构大楼等附加主题的能力。佛罗里达州和加利福尼亚州的建筑设计行业不断刷新着卓越设计的新高度。而HUUM建筑事务所的独特定位将得以利用这种极不寻常的安排以及这种安排带来的客户群体。具体事宜，未来五年将见分晓。

（皮埃路易吉·赛拉诺是旧金山湾区的执业建筑师，作家和教育家。）

Beach Road 2
海滨路2号

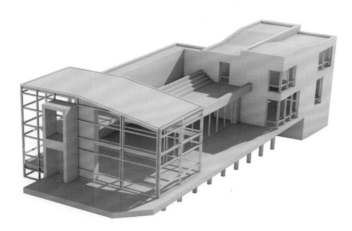

Beach Road 2 lies in a terrain of sensitive coastal sand dunes stretching along a Floridian barrier island. Responding to environmental restrictions on the shape of the building's envelope, we threaded the house between a pre-existing sea wall and the state coastal control line. The primary spaces of the house are arranged into three distinct parts varying in their material and openness. Enclosed by a taut, fretted glass envelope blurring the boundary between indoor and outdoor, the living room provides an elevated view of the waters beyond. Linking the living to the master wing, the central kitchen area is the main gathering and entertaining area; the professional-grade kitchen becomes a magnet for visitors and owners alike. Open to the east, a wall of aluminum louvers protects the room from the low western sun. All rooms look to the east, and the walls become more transparent as they march south. The bedroom wing is the most solid, with punctured windows set into stucco and colored to match the surrounding sand; seen from the adjacent beach, the house appears to be rising from the landscape. Two guestrooms sit atop the master bedroom. On the opposite end of the house is a frameless two-story living room. Entering from a sloped entry hall, visitors discover light entering the space from the underside of a glass-bottomed rooftop lap pool. Whether viewed from above or below, the compelling nature of the pool provides a visual and visceral link between the house and the surrounding waters; this is a place of nuance and discovery.

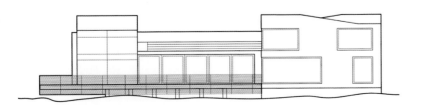
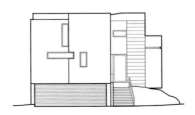
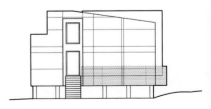

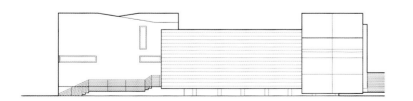
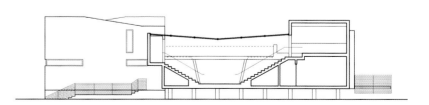

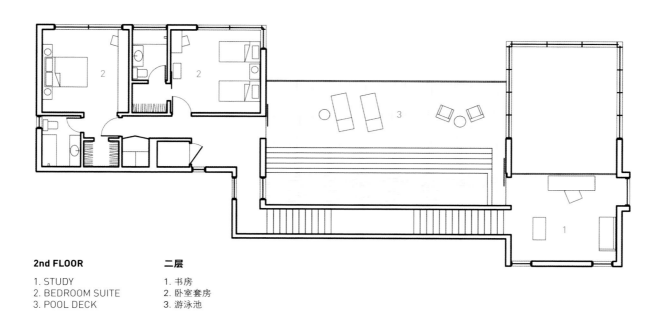

2nd FLOOR　　二层

1. STUDY　　　　　1. 书房
2. BEDROOM SUITE　2. 卧室套房
3. POOL DECK　　　3. 游泳池

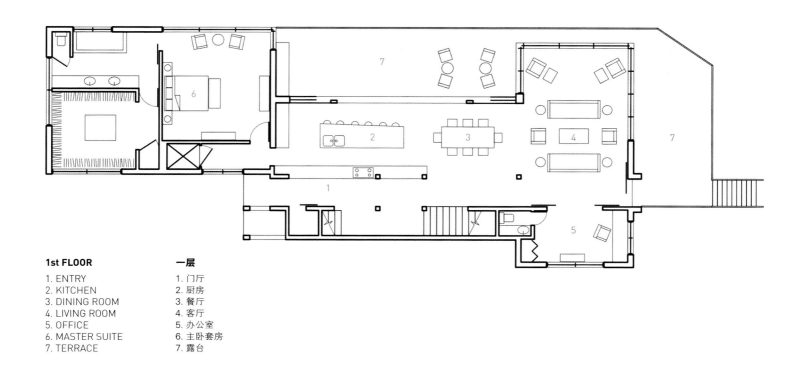

1st FLOOR　　一层

1. ENTRY　　　　　1. 门厅
2. KITCHEN　　　　2. 厨房
3. DINING ROOM　　3. 餐厅
4. LIVING ROOM　　4. 客厅
5. OFFICE　　　　　5. 办公室
6. MASTER SUITE　　6. 主卧套房
7. TERRACE　　　　7. 露台

海滨路2号位于佛罗里达州一个堰洲岛上。这里起伏的沙丘紧邻大西洋,以此为界,房屋的主体空间被分为三个不同区域,各自使用不同的建筑材料,呈现不同的开放度。格子玻璃外墙模糊了室内与室外的界限。在这里可以从较高的视角清楚地望见不断变化的水景。连接这两个部分的中央厨房是主要的聚会或娱乐区。专业级厨房对访客和住宅的主人都有着十足的吸引力。朝向东方的铝质百叶窗能够抵挡日照。所有房间都朝向东方,充分利用了这一侧风景秀丽的优势,越靠近南侧的墙壁上玻璃使用面积越大。卧室一侧的设计最为坚实,窗口嵌入颜色与周围沙滩相配的灰泥,似乎从相邻的海滩中拔地而起。两间客房位于主卧上方。房子另一侧是视野开阔的两层客厅,从略有坡度的入户大厅进入室内,访客会惊喜地沐浴在穿过玻璃底屋顶游泳池照射进来的光线之中。无论从上方还是下方观察,引人注目的透明游泳池设计都在房屋与周围水景之间发挥着视觉与内部连接的作用。这是一个充满精妙细节和惊喜发现的地方。

Big Timber
大廷伯河畔休闲住宅

At the base of the Montana mountain range known as the Crazies, this residence is a gathering place for an active family. The sole shade trees on this 8 km² ranch are the cottonwoods along the banks of the property-bisecting river. The house is sited in a clearing within a grove of these cottonwoods, revealing a clear view of the distant mountain range.

The river has the potential to dramatically flood its banks during late spring melt-off in the mountains. The house sits on a 0.76m tall porous plinth, lifting the finish floors well above the flood plane. The house has been sited among spring-fed ponds in the flood plain, providing year-round water features when the river can otherwise slow to a trickle during drought cycles.

On arrival, the house presents two distinct and separate facades, revealing itself as two interlocking objects once the visitor enters: one a two-level glass wedge, the other a one-story wooden bar. A dramatic angular box with glass walls facing the extraordinary mountains contains the living and dining room. Anchored by a tall, galvanized steel fireplace, it floats above the prairie, allowing natural drainage to move through the site. Ipe ceilings and reclaimed barn wood walls extend to the exterior, framing the rugged views and sheltering an outdoor deck. A glass-enclosed hall along the western side of the residence adds to the width of the wooden structure, recalling in scale and function the shed-covered walkways of former frontier towns.

Completion Date: 2004
Area: 334.45sqm
Location: Jupiter Island
Landscape Design: huum architects
Photography: Ken Hayden

建成时间：2004年
面积：334.45平方米
地点：佛罗里达州，朱庇特岛
景观设计：huum建筑事务所
摄影：肯·海登

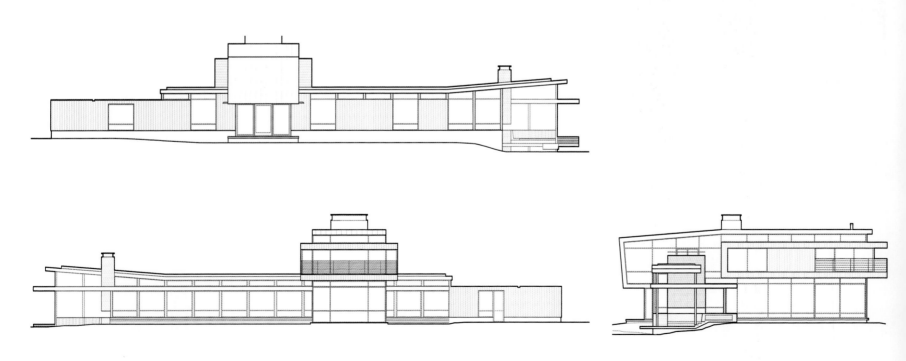

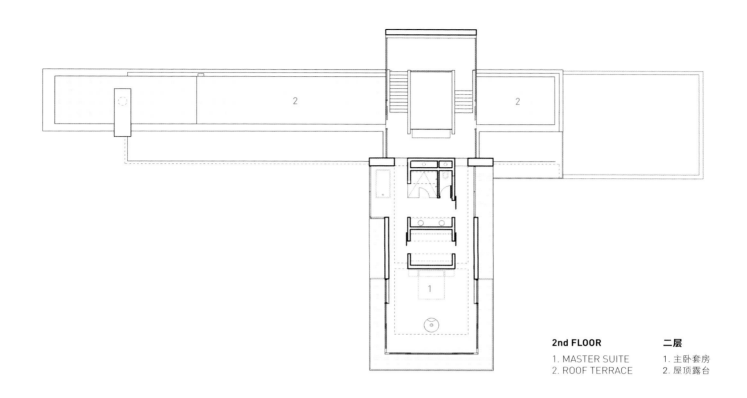

2nd FLOOR　　　二层

1. MASTER SUITE　　1. 主卧套房
2. ROOF TERRACE　　2. 屋顶露台

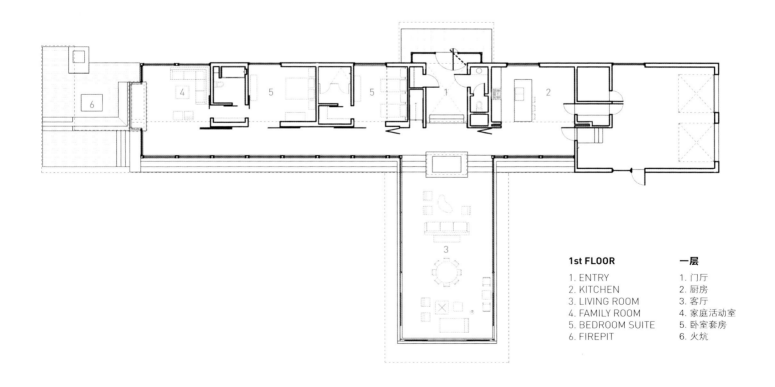

1st FLOOR　　　一层

1. ENTRY　　　　　1. 门厅
2. KITCHEN　　　　2. 厨房
3. LIVING ROOM　　3. 客厅
4. FAMILY ROOM　　4. 家庭活动室
5. BEDROOM SUITE　5. 卧室套房
6. FIREPIT　　　　6. 火炕

这处适合活跃家庭的聚会住宅位于蒙大拿山脉的山脚。这条山脉被称为"Crazies",意为"疯狂的人"。河流将住宅一分为二,这片8平方千米的牧场上仅有的遮阴树便是沿着河岸生长的杨树。项目位于杨树中间的一片空地,可以清晰地望见遥远的山脉。

初春山间的冰雪融化时,河岸有被河水冲击的风险。 设计师选择将项目建在0.76米高的多孔基座之上,建成后的地面远高于洪水高度。泉水滋养的池塘将房屋包围,形成全年可见的水景,即便在干旱时期也是如此。

项目在入口处呈现两个风格不同的独立外墙,向来访的客人展示自身作为两个交错结构的特征:一个是两层的楔形玻璃结构,另一个是一层的木质结构。朝向雄伟山脉的角框配有玻璃墙壁,造型夸张,是客厅与餐厅所在的位置。由高大的镀锌钢壁炉固定,角框悬浮在草甸之上,自然排水系统贯穿整个场地。重蚁木天花板和回收的谷仓木质墙板向室外延伸,呈现高低错落的景致,遮挡户外墙面。住宅西侧的玻璃幕墙大厅增加了木质建筑的宽度,让人想起旧时边境城镇有棚人行道的时尚和功能。

Completion Date: 2013
Area: 353.03sqm
Location: Montana
Landscape Design: Raymond Jungles
Photography: Gibeon Photography

建成时间：2013年
面积：353.03平方米
地点：蒙大拿州
景观设计：雷蒙德·江格斯
摄影：吉比恩摄影公司

Stone
斯通宅邸

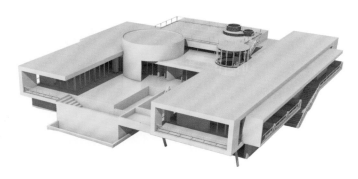

Recognizing the "green" value in renovation vs. new construction, the client, an environmental lawyer, purchased a tired single-family house on a barrier island along the Treasure Coast of Florida. Located on the narrowest part of the island, the site affords clear views of both the Atlantic Ocean as well as the Intracoastal Waterway.

The program called for a reduction in size and re-envisioning of the original 743 sqm single story home built in 1991. The new elevations display simplicity in form lacking in the original design. A dramatic staircase links the single story living level to the garage level below and the roof deck above.

Preserving the existing foundation, the original concrete shell was modified to accept new windows glazed with state of the art insulated impact glass featuring "VIEW" technology that automatically tints itself on demand throughout the day.

The walls are detailed as rain-screens clad with FSC certified wood. The west facing water feature acts as a cooling tower to supplement the geothermal HVAC system.

These environmentally friendly specifications qualify this project for LEED Silver Certification.

East Meets West
东方与西方相遇

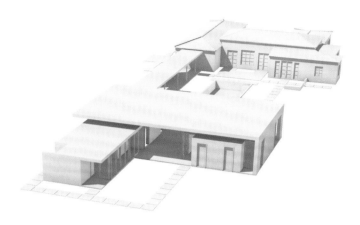

The client has collected an interesting mix of eastern art (south Asian sculptures) and western art (18th and 19th century American portraits and landscapes). Their 1935 Maurice Fazio home was too small to display the collection. They came to the design team for an expansion. The program expectations doubled the existing square footage. Adding a giant wing would overwhelm and diminish the charm and identity of the existing house. Instead designers looked to the un-built lot to the east.

The solution: build a new pavilion informed and inspired by the scale and characteristics of the original house. A centerline through the living rooms established the pool location. This lap pool, along with a covered walkway, creates the visual and physical connection between the two structures. Where the covered walkway touches the old and the new has been carefully considered.

Using the existing building façade as a guide for the new pavilion's design, the team mirrored the scale and massing of the existing building onto our new structure on the eastern half of the site, beginning a literal and metaphoric architectural dialogue between east and west.

Though examples of each collection are displayed throughout the buildings, the Asian collection resides primarily in the east and the early American art is on display mainly in the west. The original house is used for private functions, the new pavilion is used for entertaining and repose--work in the west, entertain in the east.

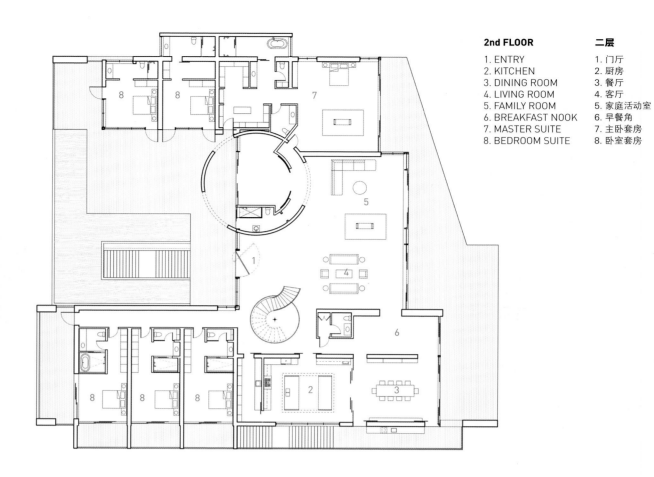

2nd FLOOR　　　　　二层
1. ENTRY　　　　　　1. 门厅
2. KITCHEN　　　　　2. 厨房
3. DINING ROOM　　　3. 餐厅
4. LIVING ROOM　　　 4. 客厅
5. FAMILY ROOM　　　5. 家庭活动室
6. BREAKFAST NOOK　6. 早餐角
7. MASTER SUITE　　　7. 主卧套房
8. BEDROOM SUITE　　8. 卧室套房

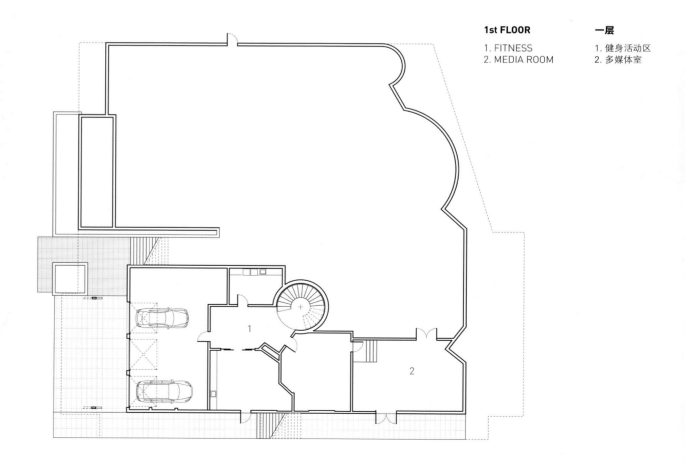

1st FLOOR　　　　　一层
1. FITNESS　　　　　　1. 健身活动区
2. MEDIA ROOM　　　　2. 多媒体室

Completion Date: 2017
Area: 743.22sqm
Location: Jupiter Island, Florida
Landscape: Gentile, Glas, Holloway and O'Mahoney

建成时间：2017年
面积：743.22平方米
地点：佛罗里达州，朱庇特岛
景观设计：Gentile, Glas, Holloway and O'Mahoney公司

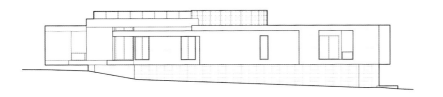
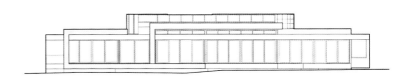

本项目的客户是一位环境律师。他在意识到与新建筑相比，改造建筑的"绿色"价值之后，购买了佛罗里达州宝藏海岸沿线一个堰洲岛上的房屋。房屋所在地位于岛上最狭窄的区域，这里可以清晰地看到大西洋以及近岸内航道的景象。

项目要求缩减房屋的尺寸，对这栋建于 1991 年占地 743 平方米的原一层建筑进行重新规划。新的立面设计展现出原建筑所不具备的简洁形式。夸张的楼梯设计将生活区与位置较低的车库以及位置较高的屋顶平台相连。

设计团队保留现有地基，将原有的混凝土外墙加以调整；增加了应用先进隔热技术"VIEW"的新玻璃窗，这种技术可以根据一天之中的不同需求进行自动调节。

墙面增添了用 FSC 认证木料制作的雨幕。朝向西方的水景充当冷却塔，为地热暖通空调系统提供补充。

这些环保规范使得项目获得了 LEED 银质认证。

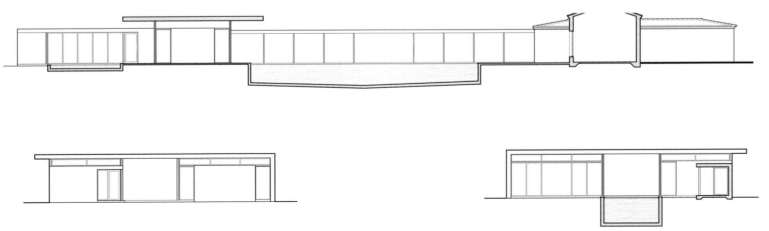

本案中的客户拥有一组有趣的东方艺术品（南亚雕塑）和西方艺术品（18、19世纪美国肖像画和风景画）。客户建于1935年的家已经没有足够的空间展示这些艺术品，因而带着扩建的想法找到设计师。项目预期将原有面积扩大一倍，增添附加结构会削减原有建筑的风格和个性，设计团队选择在未建的东侧结构上着手。

设计方案：参考原建筑的规格和特点规划新结构，利用穿过客厅的中心线确定游泳池的位置，游泳池与有棚通道构成新旧建筑之间的视觉与实体连接，有棚通道与新旧结构相连的位置均经过了仔细的思考验证。

设计师以现有建筑外墙为指导，进行新结构的设计。将现有建筑的规格和体量投射到场地东侧的新建筑上，实现东方与西方在字面含义与比喻层面上的对话。

建成后，亚洲藏品主要集中在新建筑的东侧，早期美洲艺术品则主要位于新建筑的西侧。原有建筑如今主要用于私人功能，新建筑则用于娱乐和休息——工作在西方，娱乐在东方。

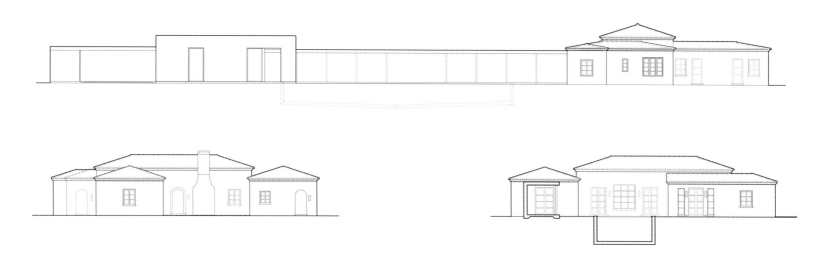

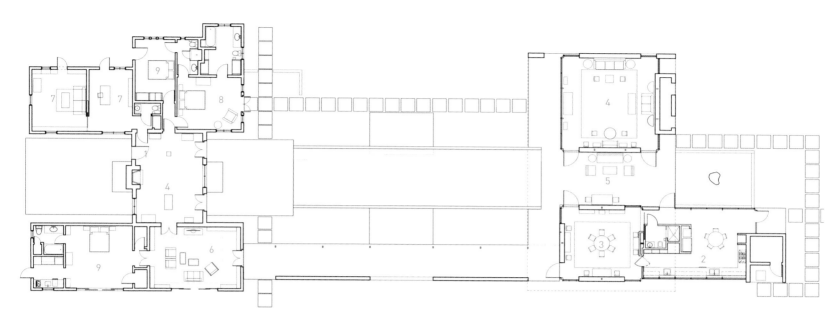

1st FLOOR

1. ENTRY
2. KITCHEN
3. DINING ROOM
4. LIVING ROOM
5. BREAKFAST NOOK
6. MUSIC ROOM
7. OFFICE
8. MASTER SUITE
9. BEDROOM SUITE

一层

1. 门厅
2. 厨房
3. 餐厅
4. 客厅
5. 早餐角落
6. 音乐室
7. 办公室
8. 主卧套房
9. 卧室套房

Completion Date: 2014
Area: 353.03sqm
Location: Jupiter Island, Florida
Landscape Design: Innocenti and Webel
Photography: Allan Carlisle Photography

建成时间：2014年
面积：353.03平方米
地点：佛罗里达州，朱庇特岛
景观设计：Innocenti and Webel设计公司
摄影：艾伦·卡莱尔摄影公司

Paseo Miramar
帕萨奥·米拉玛

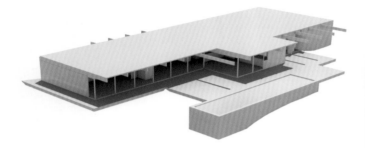

Originally built in 1956, Richard Neutra's Troxell Residence perched high above the Pacific Coast Highway in the Los Angeles area known as Pacific Palisades, underwent a rigorous renovation and expansion to 278 sqm. Untouched since its construction, the house was in need of a total rehabilitation, but with a light touch.

Following cues from the original post and beam structure, interior and exterior finishes of wood, stone, plaster and glass were applied in a simple refined technique to emphasize the play of space, light and views.

A pile supported pool and deck, included in the original plans but never executed, now clings to the natively planted hillside, defining the limit of expanded outdoor living space and void below. Three bedrooms are located off a common corridor facing west to the hillside, making for consistent, calm morning light. A seamless addition to the master bedroom and master bath projects the house further out into space - reemphasizing the overall horizontal composition of the building. From most rooms, including (note the space) the open plan dining and sitting areas large expanses of fixed and sliding glass create framed panoramic views that include downtown Los Angeles to the east and Catalina Island to the west.

Some of the included photos were the last taken by the famed photographer Julius Shulman. 50 years after Mr. Shulman took his first set of images of the project for Art and Architecture Magazine, we invited him back to reshoot the project again. The addition/renovation was completed in 2006.

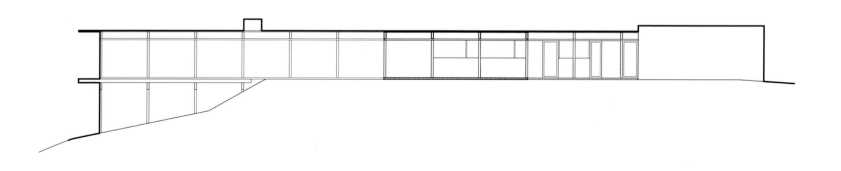

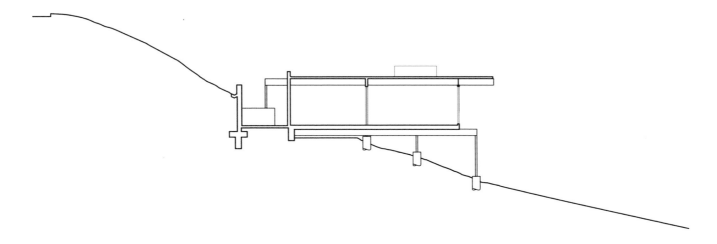

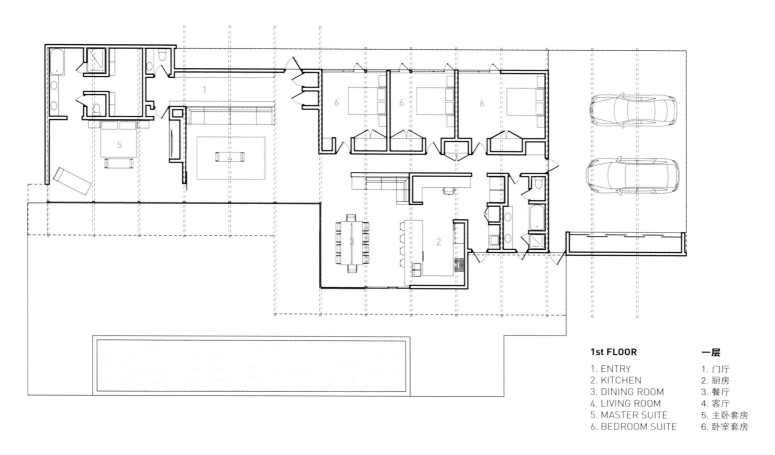

1st FLOOR   一层

1. ENTRY   1. 门厅
2. KITCHEN   2. 厨房
3. DINING ROOM   3. 餐厅
4. LIVING ROOM   4. 客厅
5. MASTER SUITE   5. 主卧套房
6. BEDROOM SUITE   6. 卧室套房

始建于 1956 年，理查德·纽佐尔的特洛塞尔宅邸坐落在洛杉矶的太平洋海岸，被称为"太平洋帕利塞兹"，经历了严密的改造和扩张工程后总面积达到 278 平方米。由于建成之后从未进行修缮，这处住宅确实需要一番彻底翻修，但对修缮力度又应有所把握。

参考原有梁柱结构，室内外空间采用了风格简约精致的木质、石质、石膏和玻璃表面材料，以突出空间、光线和视角的变化关系。

原始计划中提出但却从未实施的一个以地桩为基座的游泳池和地板如今建在长满本地植被的山坡上，划分出室外休闲空间和留白的界线。三间卧室位于朝向西侧山丘的公共走廊一侧，清晨时段光照持续充足。主卧室和主卧浴室的增建结构自然流畅，使得空间更为延伸，进一步突出项目整体的水平结构。包括开放式餐厅和客厅在内的大部分房间都安装了大面积的固定和滑动玻璃窗，窗外是东起洛杉矶市区西至卡特琳娜岛的秀丽全景。

项目中选用的一些照片为摄影师朱利叶斯·舒尔曼的作品。50 年前，朱利叶斯·舒尔曼受《艺术与建筑杂志》邀请为这里拍摄了第一组照片。如今他再次受邀为这个项目拍摄照片。扩建改造项目于 2008 年完工。

Completion Date: 2004
Area: 334.45sqm
Location: Pacific Palisades, California
Landscape Design: Judy Kameon
Interior Design: Brad Dunning
Photography: Julius Shulman, Nick Springett

建成时间：2004年
面积：334.45平方米
地点：加利福尼亚州，太平洋帕利塞兹
景观设计：朱迪·柯米恩
室内设计：布拉德·邓宁
摄影：朱利叶斯·舒尔曼，尼克·斯伯格蒂

Baker's Bay
贝克湾

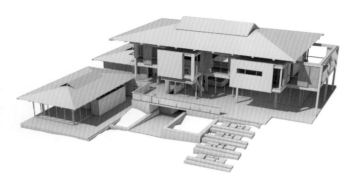

Baker's Bay Golf & Ocean Club is a private resort community in Northeastern Bahamas. The development is located on Great Guana Cay situated between the Sea of Abaco and the Atlantic Ocean. The 2.37 km² project is a resort development of 385 homes. Developed by Discovery Land Company, the project includes a Tom Fazio-designed 18 hole golf course, 200-slip marina, the Marina Village resort area, and a private club that includes beach club and spa. The client purchased a 3,000 sqm lot on the Atlantic side of the island. At the time when the work began no other houses were under construction along this strip of ironshore beach.

On the third attempt at working within the ridged architectural guidelines, we were able to get authorization to begin construction. The earlier solutions were judged too modern, so the designers took characteristics of the pattern book provided by the developer and re-configured them into an architecture that would match the expectations and lifestyle of our client.

Working with renowned Miami landscape architect Raymond Jungles, the design team created a twin dune, parallel to the untouchable original, to elevate the main level to engage the breaking waves. The tall ceiling of the two-story living space creates a natural ventilating source exhausting the warmer air in favor of the cool. Aware of the strong breezes coming off of the ocean at night, a large outdoor living space uses the walls of the house as a windbreak.

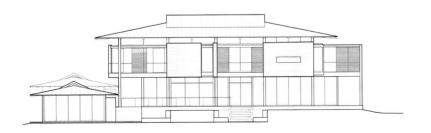

Area: 427.35sqm
Location: Guana Cay, Bahamas
Landscape Design: Raymond Jungles

面积：427.35平方米
地点：巴哈马，大瓜纳岛
景观设计：雷蒙德·江格斯

贝克湾高尔夫与海洋俱乐部是巴哈马群岛东北部的一处私人度假区。项目位于阿巴科海和大西洋之间的大瓜纳岛上，占地2.37平方千米，包含385间房屋。项目由Discovery Land公司开发；包含了汤姆·法齐奥设计的18洞高尔夫球场，200个泊位的码头，Marina Village度假区，一个配备了沙滩俱乐部和SPA的私人俱乐部。客户买下了岛上靠近大西洋一侧3,000平方米的土地。项目开始动工之时，这处海滩上还没有其他正在建设中的房子。

设计方案经过三次修改后终于获得批准。之前的设计方案被批评过于现代，所以设计师利用开发商提供的项目任务书中描述的特征，绘出了建筑设计方案，最终达到了客户的期望，符合客户的生活方式。

设计团队与迈阿密著名景观设计师雷蒙德·江格斯合作，打造出一对与原有景观平行的沙丘，提高了项目主体高度，以应对海浪。两层活动空间的挑高天花板形成自然通风，排出暖热空气，引进凉爽气流。考虑到夜间的猛烈海风，设计师还利用房屋的墙壁，构建了一个防风的大型户外活动空间。

2nd FLOOR   二层

1. LIVING ROOM   1. 客厅
2. MASTER SUITE   2. 主卧套房
3. BEDROOM SUITE   3. 卧室套房
4. OPEN TO BELOW   4. 天井
5. BALCONY   5. 阳台

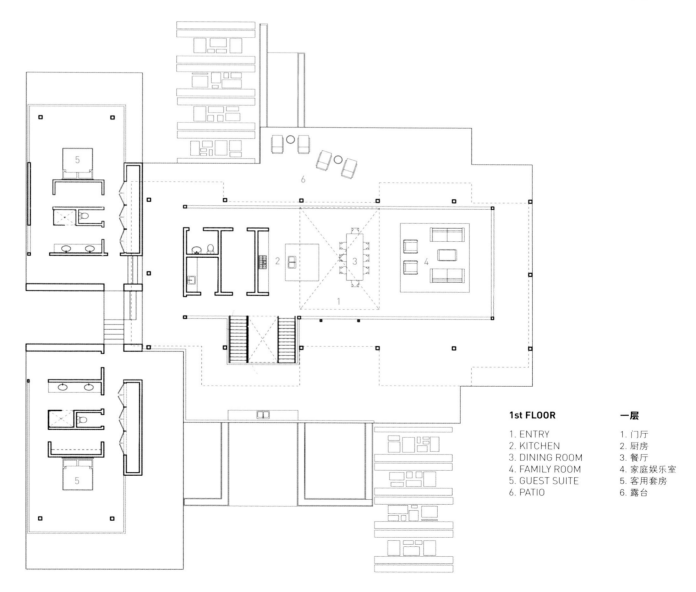

1st FLOOR   一层

1. ENTRY   1. 门厅
2. KITCHEN   2. 厨房
3. DINING ROOM   3. 餐厅
4. FAMILY ROOM   4. 家庭娱乐室
5. GUEST SUITE   5. 客用套房
6. PATIO   6. 露台

Grenville
格伦维尔

"Can you design for us a house that has no straight walls?" began the interview to become the architects for this unexpected commission. A young cosmopolitan couple living in Europe and with the resources to plan for the future was looking to build a house along the Intracoastal Waterway. It would be used only for holidays until their teenage children were on their own. But the parents made it clear from the start that this would always be a "play house", nothing about it needed to be formal. This was intended to be a welcoming family destination for years to come, able to comfortably shelter multiple generations looking forward to spending time together.

The thousands-of-sqm site consisted of two flat levels separated by a 3m slope. This unnatural configuration resulted from the site of the previous house built in the 1940's, torn down 50 years later. A single yellow flowering tree remained from the original landscaping. Learning of its rare existence the designers decided to allow its location to define the layout of the house. Basically the team built the house around the tree.

A series of pavilions are connected to a ribbon shaped hallway curving around the branches of the tree. Each pavilion houses a specific program: sleeping, living, dining etc. Unique roof shapes define each of the pavilions as well. There is a daytime side of the house where the rooms look out to the river. On the opposite end of the house is the nighttime room where a 100-year-old ficus tree is the focus visible through the curving glass facade.

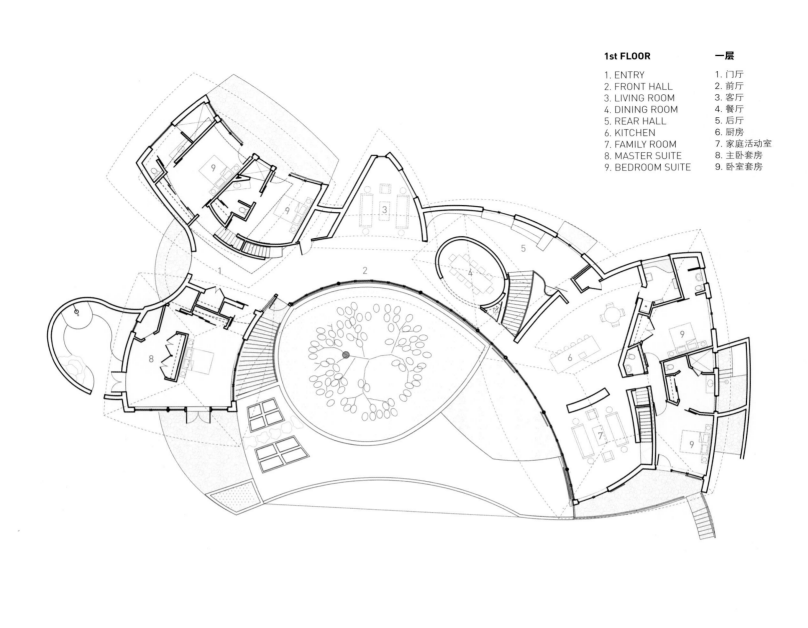

1st FLOOR 一层

1. ENTRY   1. 门厅
2. FRONT HALL   2. 前厅
3. LIVING ROOM   3. 客厅
4. DINING ROOM   4. 餐厅
5. REAR HALL   5. 后厅
6. KITCHEN   6. 厨房
7. FAMILY ROOM   7. 家庭活动室
8. MASTER SUITE   8. 主卧套房
9. BEDROOM SUITE   9. 卧室套房

项目答辩一开始,设计师就遇到了这样的提问,"可以为我们设计一间没有直墙的房子吗?"这是一次让人意想不到的委托。一对居住在欧洲,崇尚世界主义的年轻夫妇带着对未来进行规划的资料找到设计团队,希望沿着近岸内航道建造房子。到他们十几岁的孩子长大成人之前,这间住宅将仅在度假时使用。

这对父母从一开始就明确表示,这里将一直是"休闲玩耍之地",完全不需要有正式的感觉。未来的几年时间里,这里将成为热闹的家庭活动场所,是几代人共度家庭时光的舒适港湾。

这块占地数千平方米的场地由两块平坦的部分组成,中间有3米的斜坡相隔。20世纪40年代于此地建造的房屋在使用50年后拆除,形成了这种不自然的空间组成。地块上有一棵开黄花的榕树被保留了下来。了解到这棵树的稀有之后,设计团队决定在它的位置基础上进行房屋的布局设计,也就是围绕这棵树建造房屋。

带状走廊以流线型围绕树枝,连接一系列建筑结构。每个结构单元容纳一些特定功能:卧室、客厅、餐厅等。独特的屋顶形状也将各个单元区分开来。房屋有一侧朝向河流,用于日间活动。相反一侧的空间则以夜间活动为主,透过曲面玻璃墙能够从各个角度观察到百岁榕树的样子。

Completion Date: 2014
Area: 826.84sqm
Location: Jupiter Island, Florida
Landscape Design: Innocenti and Webel
Photography: Tom Winter

建成时间：2014年
面积：826.84平方米
地点：佛罗里达州，朱庇特岛
景观设计：Innocenti and Webel设计公司
摄影：汤姆·温特

Pine School
潘恩学校

Intending to relocate to a less urban area, the trustees of this private day school chose to locate their new campus on 0.69 km² of untouched sand dunes 4.8 km off the Atlantic Coast. Entitlement studies revealed that the property was environmentally sensitive, reducing the buildable area to just under 0.08 km². On this blank canvas the design team were privileged to envision a new academic village.

Education requires an elusive dose of order and chaos, predictability and improvisation in order to succeed. The design intends to provide just such a backdrop. The program for this school called for classrooms and labs for group learning, a library for individual and group study, and places for assembly, performance, athletics and relaxation. It is an academic campus where groups emerge, are disbanded and regroup: a community for learning, growing and discovering. The campus design is a functional and formal response to a specific educational vision and was programmed for both active and continuous expansion and inward change. The buildings are specifically located facing each other to create an inwardly focused lawn with simultaneously open yet enclosed spaces. A repeatable two-story module, with each containing seven classrooms, covered circulation, lockers and toilets, became the prototype for construction and future planning. One new module is scheduled to be added to the campus every five years for the next twenty years.

This structure is designed to promotes education and profoundly engages the specifics of its location: "softly touching" the environment and re-defining itself through the introduction of children and teachers.

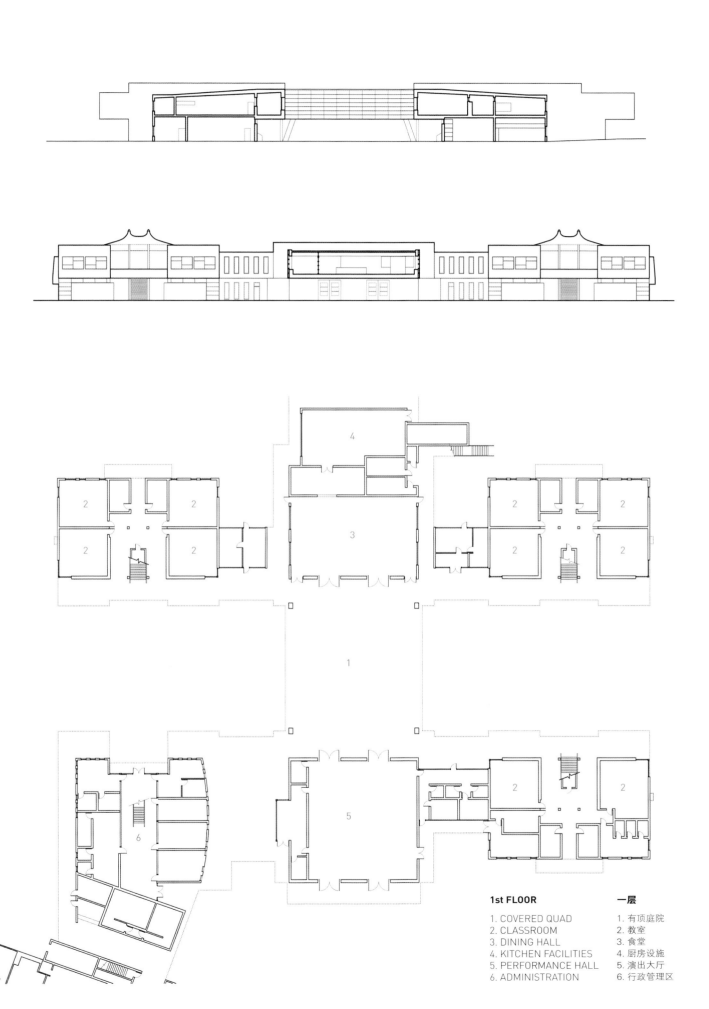

1st FLOOR　　　　　一层

1. COVERED QUAD　　1. 有顶庭院
2. CLASSROOM　　　2. 教室
3. DINING HALL　　　 3. 食堂
4. KITCHEN FACILITIES　4. 厨房设施
5. PERFORMANCE HALL　5. 演出大厅
6. ADMINISTRATION　　6. 行政管理区

097

这间私立走读学校的受托人计划将学校迁至稍微远离市中心的地方，最终选择了距离大西洋海岸 4.8 千米，一块占地 0.69 平方千米的原始沙丘区作为新校区的所在地。研究表明由于校园建筑对环境有较高的敏感度，适合建造的区域只有不到 0.08 平方千米。设计团队就在这块空白的土地上展开关于新校园的畅想。

教育成功之前，秩序与混乱，可预见性与即兴发挥大量共存。该设计方案力争为这样的反映提供适宜环境。项目需求包括适合小组学习的多个教室和实验室，进行个人学习及小组研究的一间图书馆，以及集会、表演、运动和休闲场地。这里是人群集合、解散并再集合的校园：一个供人们学习、成长和发现的社区。

校园设计是对特定教育愿景的功能与形式的回应，应该具备对积极、连续扩张与内在改变的适应性。项目中的建筑朝向彼此相对，围成相对封闭的草坪，形成既开放又封闭的独特空间。可重复的两层结构单元每个包含 7 间教室，有棚走廊、储物柜和厕所是施工和未来规划的标准配置。项目还计划在未来 20 年中，每 5 年在校园中增加一个新单元。

工程的目的是推广教育，更加深入地利用项目地点的特点，"以轻柔的方式触及"环境，通过学生与教师的作用完成自身的再定义。

Completion Date: 2007
Area: 9,011.59sqm
Location: Hobe Sound, Florida
Landscape Design: Lucido Associates
Photography: Ken Hayden

建成时间：2007年
面积：9,011.59平方米
地点：佛罗里达州，霍布桑德
景观设计：卢奇多联合事务所
摄影：肯·海登

Tasting Room
品酒屋

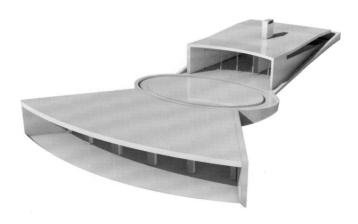

This 0.49 km² farm in the Umbrian countryside of Northern Italy had been dormant for 40 years prior to its acquisition in 2000. The new owners from America considered many options to make the land self-sustaining before they settled on winemaking.

In support of that decision, this structure is intended to house the farm equipment, but more importantly to be a gathering place where the wine can be stored and sampled. This oval space has a continuous skylight at the intersection of roof and wall that creates a constantly shifting border of shadow and light. The atmosphere is intended to be sophisticated yet simple and rustic. In addition, a small apartment unit houses several of the seasonal workers.

Stringent local regulations restricted the architectural characteristics of any new construction. Only a third of the exterior walls could be above grade. 50% of those exposed walls had to be clad with local stone. In lieu of the proscribed clay roof tile the designers choose to plant the roof in local grasses and flowering plants and appealed to the local building officials for relief.

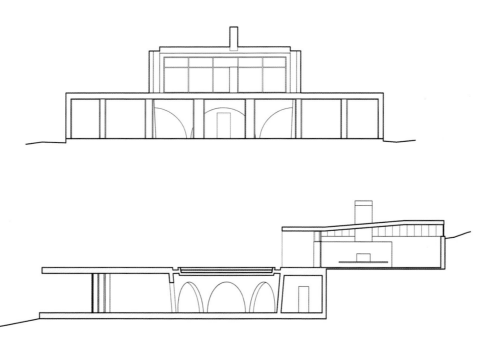

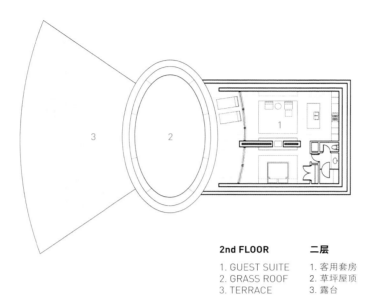

2nd FLOOR 二层

1. GUEST SUITE   1. 客用套房
2. GRASS ROOF   2. 草坪屋顶
3. TERRACE   3. 露台

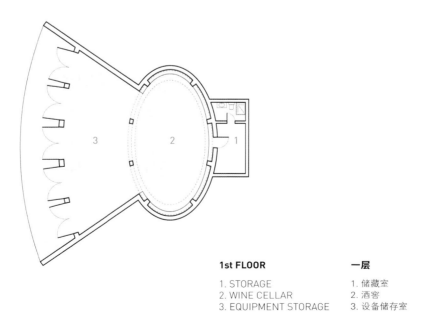

1st FLOOR 一层

1. STORAGE   1. 储藏室
2. WINE CELLAR   2. 酒窖
3. EQUIPMENT STORAGE 3. 设备储存室

Completion Date: Approximately 2018
Area: 260.13sqm
Location: Umbria, Italy
Landscape Design: huum architects

建成时间：约2018年
面积：260.13平方米
地点：意大利，温布利亚
景观设计：huum建筑事务所

在 2000 年被收购之前，意大利北部温布利亚乡间的这座 0.49 平方千米的农场已经沉睡了 40 年的时间。来自美国的农场新主人考虑了让土地自给自足的诸多选项之后，最终选择了酿酒。

为了配合这一决定，需要增设农场建筑，一方面用于存放农场设备，更重要的是建造一个对红酒进行储存和取样的人群聚集空间。椭圆形空间的屋顶与墙壁交汇处设有连续天窗，形成光与影不断变化的效果。设计力争打造具有格调又简单朴素的氛围。另有一座小型公寓为季节性工人提供住宿。

新工程的建筑设计受到严格的地方法规的限制。仅有三分之一的外墙可以露出地面。50% 的裸露墙面需要使用本地石材覆面。由于禁止使用屋顶瓦片，作为替代，设计师选择在屋顶采用本地草种和种植开花植物，并呼吁当地建筑官员提供帮助。

River Road
河畔别墅

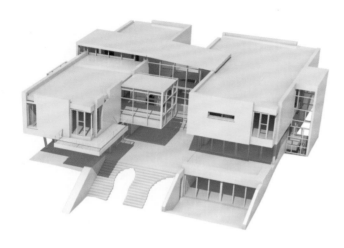

Surrounded by intracoastal waterways, the sea breeze, warm sun and panoramic views of the St. Lucie River grace the 8,000 sqm site of this landmark home. Continuing down the driveway one discovers the house perched on the edge of the sloping site; it is arranged to screen neighboring homes while creating a sense of openness and transparency. Sitting on a concrete structured base are two mirrored elements, similar yet unique, sheathed in Minnesota limestone with exposed corners of natural ipe. A dramatic glass angular volume links these wings with stairs at either end. The north wing contains the living room, with the master bedroom and guest suite above. The southern wing houses the kitchen and family room, with two children's bedrooms above them. Within the glass wedge is a bridge connecting the two wings with a dining space below. The structure is a series of expansive, intimate courtyards and cantilevered decks, which architectonically echo the volumes of the interior. The house comprises a series of distinct yet interlocking rectilinear volumes joined by a transparent longitudinal axis. Positive volumes are balanced atop negative, creating heightened tension and intrigue. A glass-skinned throughway connects and distinguishes the two dominant volumes. Interior spaces are carefully orchestrated to conceal and reveal dramatic views to the water. Its position on the site, imaginative geometry, and creative use of glass, limestone and steel give great strength to this waterfront home.

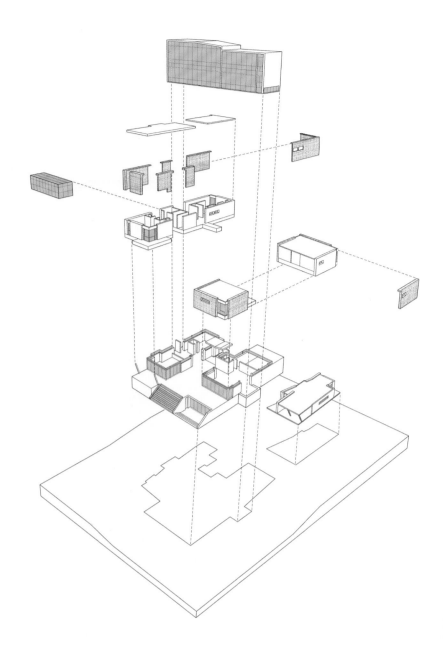

122

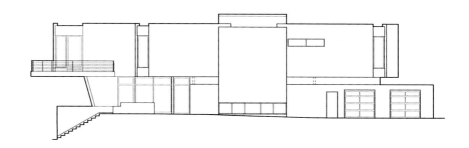

 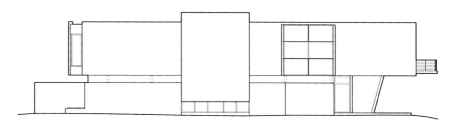

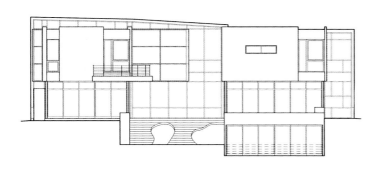 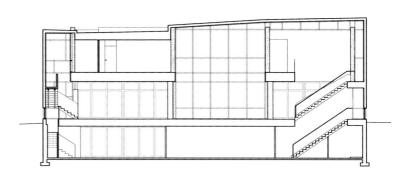

123

2ed FLOOR   二层

1. MASTER SUITE   1. 主卧套房
2. OFFICE   2. 办公室
3. STUDY   3. 书房
4. GUEST SUITE   4. 客卧套房
5. OPEN TO BELOW   5. 天井
6. TV ROOM   6. 电视房
7. BEDROOM SUITE   7. 卧室套房

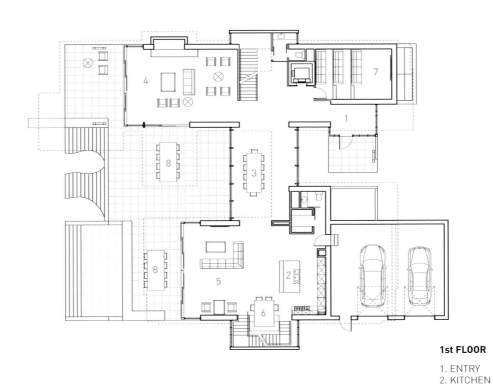

1st FLOOR   一层

1. ENTRY   1. 门厅
2. KITCHEN   2. 厨房
3. DINING ROOM   3. 餐厅
4. LIVING ROOM   4. 客厅
5. FAMILY ROOM   5. 家庭活动室
6. BREAKFAST NOOK   6. 早餐角
7. MEDIA ROOM   7. 多媒体室
8. COVERED PATIO   8. 有顶露台

项目位于佛罗里达州东海岸被称为"塞维利亚角"的半岛上一处山脉顶部。沿着曲折的道路来到山顶,客房是沿街映入眼帘的第一部分建筑。这里被近岸内航道围绕,海风轻拂,阳光普照,圣露西河的美景尽收眼底。河流上游便是这栋占地8,000平方米的标志性住宅。项目坐落于倾斜地势的边上,可以沿公路驾车到达。建筑的设计安排遮蔽了相邻的房屋,展现出开放性和通透性。项目采用混凝土结构基座,地上为两个相似却各有特色的对应结构,外墙使用明尼苏达石灰岩,边角处为裸露的天然重蚁木。夸张的玻璃结构连接两个对应结构,两侧各有楼梯。北侧结构为客厅,楼上是主卧和客房。南侧结构为厨房和家庭活动室,楼上是两间儿童房。连接两侧的玻璃楔形结构也是一处就餐空间。整个建筑是一系列兼具扩张性与私密性的庭院与悬吊平台的组合,在建筑层面上与室内空间相呼应。项目由一系列独特而相互关联的直线结构组成,一条透明的纵轴将其连接,正负结构相互平衡,形成更强的张力,玻璃通道连接两个主要结构,并对二者的特征加以突出。住户可以根据自己的需要和情绪,以休闲惬意或直接高效的方式通行。室内空间的安排巧妙地呈现秀美的河景。优越的位置,富有想象力的几何造型,玻璃、钢铁和石灰石材质的精妙使用都使这间河畔别墅更加让人印象深刻。

Completion Date: 2004
Area: 334.45sqm
Location: Sewell's Point, Florida
Landscape Design: huum architects
Photography: Steven Brooks

建成时间：2004年
面积：334.45平方米
地点：佛罗里达州，塞维利亚角
景观设计：huum建筑事务所
摄影：史蒂文·布鲁克斯

One for One
One for One住宅

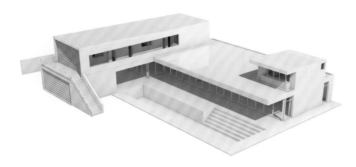

This house on Gomez Road simply and elegantly links the elements of site, material and need to create a distinct home for its residents. Challenged to foster a sense of distinction, assimilation, connection and introspection, the solution integrated existing site elements, a spare materials palette and the elegant union of simple yet bold shapes. Despite the modest 315 sqm print the two-story L-shaped structure feels larger, playing with compression and release, resulting in a comfortable external insulation from neighbors and internal openness. The first floor, with its continuous wall of operable glass, houses the day-to-day living spaces while the second floor acts as a separate, independent guesthouse. The poured-in-place concrete structure is refined and robust, built to withstand hurricane-force winds while providing an elegant framework for living. Clerestory windows are introduced to the closed side of the house to augment the sense of space, expanding the interior into a "borrowed landscape." Outside, the use of a grass roof and the strategic placement of a specimen of ficus create both visual interest and focal points in the landscape, extending the limited views. Comprised of native grasses, the accessible roof acts to enhance air conditioning efficiencies and reduce heat island effects. Within the abode, a separate external stairway adds a compelling visual vertical counterpoint to the dominance of horizontal planes, while functionally providing a private entry to the second floor guest suite.

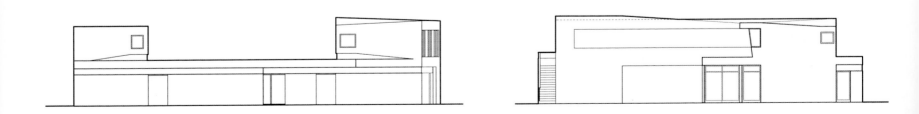

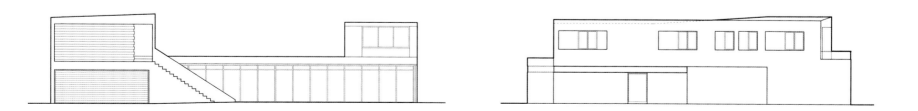

135

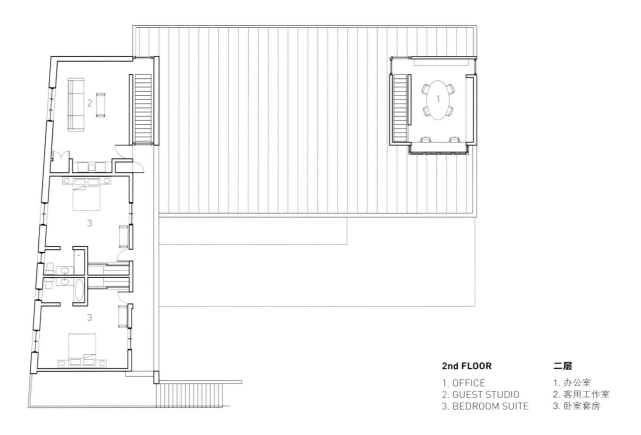

2nd FLOOR   二层

1. OFFICE   1. 办公室
2. GUEST STUDIO   2. 客用工作室
3. BEDROOM SUITE   3. 卧室套房

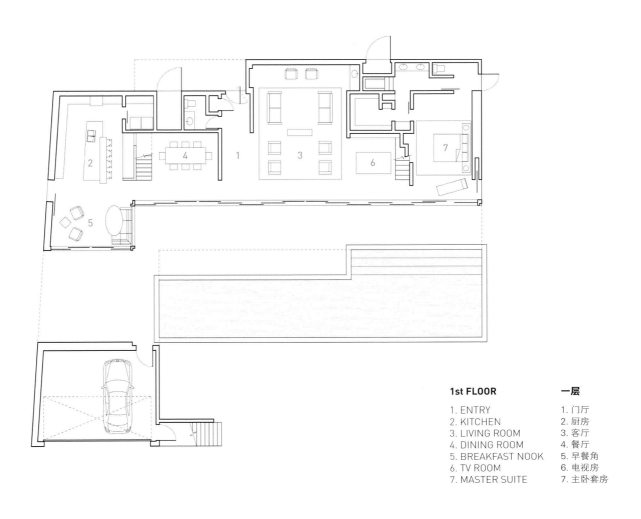

1st FLOOR   一层

1. ENTRY   1. 门厅
2. KITCHEN   2. 厨房
3. LIVING ROOM   3. 客厅
4. DINING ROOM   4. 餐厅
5. BREAKFAST NOOK   5. 早餐角
6. TV ROOM   6. 电视房
7. MASTER SUITE   7. 主卧套房

这间位于戈麦斯路上的住宅将场地周边元素、材料与客户需求以简洁而优雅的方式结合，为住户创造了一个独特的居住环境。设计团队接受这一挑战并充分享受这次机遇。 独特感与同化感并重，连通感与独立感并行，设计方案集合了现有场地元素，备用材料选项与简洁、大胆造型的优雅结合。尽管项目只有315平方米，两层的L形结构充分利用了压缩与释放的关系，给人十分宽敞的感觉，同时营造出与外部的隔离感以及内部的开放感。一楼的日常生活空间由连续可操作玻璃墙包围，二楼则作为独立的客房使用。现场浇筑的混凝土结构工艺精制、坚固，能够承受飓风等级的强风，提供优雅生活空间。天窗设计位于房屋封闭的一侧，增加空间感，将内部空间扩大成为"借来的风景"。室外草坪屋顶的使用以及一株勒颈无花果的战略性设置组成视觉兴趣点以及延伸至视线之外的景观焦点。本地草种组成的开放屋顶用于提高空调效率，降低热岛效应。郁郁葱葱的树木起缓冲作用，绿色的屏障保护住户不受干扰，同时增强隐私性与静谧感。住所内的独立外部楼梯在以水平为主的设计上增加了垂直的视觉元素，同时也是到达二楼客用套房的私密入口。

Completion Date: 2000
Area: 315sqm
Location: Jupiter Island, Florida
Landscape Design: Innocenti and Webel
Photography: Robin Hill

建成时间：2000年
面积：315平方米
地点：佛罗里达州，朱庇特岛
景观设计：Innocenti and Webel设计公司
摄影：罗宾·希尔

Hutchinson Island
哈钦森岛

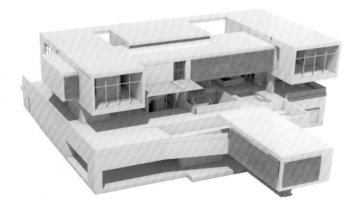

This house is a reward for a successful business career, to be used as the base camp for new ventures and a more relaxed lifestyle. Located on a double lot on a Floridian barrier island, the site provides over 76.2m of beach frontage. A major disadvantage of the site is a dune line blocking the view of the ocean from the ground level. State flood level designations also limit the usability of this first level to storage and mechanical spaces. The house will be used year round by the husband and wife and a teenage daughter. Great interest has been given to the entry sequence that rises from the ground level to the living level, established 4.27m above the ground. For the visitor it is a grand staircase rising between water features that sparks the initial interest - spilling and capturing water around you until reaching the front door and a view of the Atlantic. For the owners, a glass enclosed circular stair with an elevator at its center provides access to the living level or the bedroom level above. Taking advantage of the wide lot and with an interest in providing a view of the beach for every space, the house is intentionally designed as primarily only one room wide. Like the "Erik" project, the design is organized into three interlocking bays connected by a continuous hallway running parallel to the beach. A two story living-dining space occupies the middle of the house with the more private spaces on either end. The three level design provides differing amounts of solidity. The lower storage is primarily solid acting as a base for the living level almost entirely enclosed by glass. The upper bedroom level is a combination of the two.

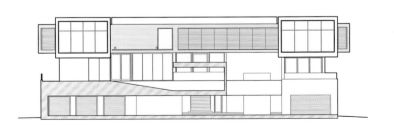

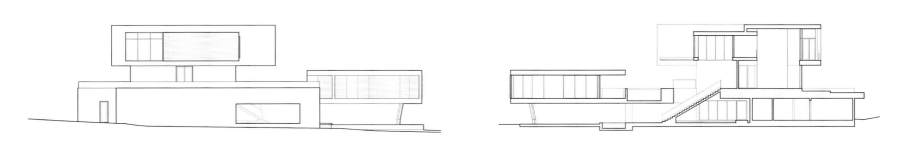

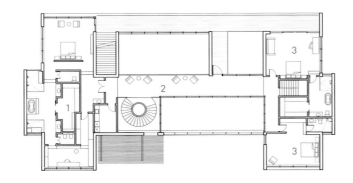

3rd FLOOR   三层

1. MASTER SUITE   1. 主卧套房
2. SITTING ROOM   2. 客厅
3. BEDROOM SUITE   3. 卧室套房

2nd FLOOR   二层

1. ENTRY STAIR   1. 门厅
2. KITCHEN   2. 厨房
3. DINING ROOM   3. 餐厅
4. LIVING ROOM   4. 客厅
5. OFFICE   5. 办公室
6. GUEST SUITE   6. 客用套房
7. TERRACE   7. 露台

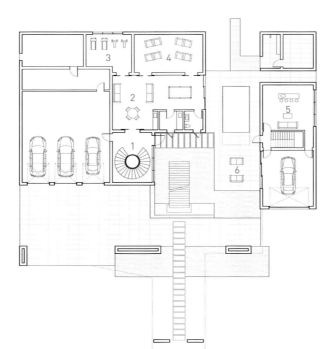

1st FLOOR   一层

1. ENTRY   1. 门厅
2. GAME ROOM   2. 游戏室
3. MEDIA ROOM   3. 多媒体室
4. FITNESS   4. 健身活动区
5. CIGAR ROOM   5. 雪茄室
6. COVERED PATIO   6. 有顶露台

这间住宅是对一段成功的职业生涯的奖励,被用作新企业的大本营,用于享受更为放松的生活方式。项目位于佛罗里达州的一个堰洲岛上,坐拥76.2米的秀美海滩。然而场地的一个主要缺点是沙丘遮挡了一楼的海景视线。美国国家洪水警戒线的标识同样使得二楼的可用性受到限制,只能做储藏室和机械室使用。这间住宅将供一对夫妇以及他们十几岁的女儿全年居住使用。设计师对入口处的设计倾注了极大的兴趣,门厅的高度为4.27米。迎接来访客人的是一个宏伟的楼梯设计,从水景之中拔地而起,沿楼梯向上一路有活泼的水花陪伴,最终到达正门,呈现在眼前的是一抹大西洋美景。玻璃围挡的环形楼梯与中央的升降电梯方便房屋主人来到上层的生活区或卧室。设计师充分利用场地的宽度优势,将房屋有意设计成细长的、约一个房间宽度的形态,打造全方位的海滩视角。与"埃里克"项目类似,设计师将项目分割成三个相互关联的部分,彼此之间借助与海滩平行的连续走廊相连。两层的客厅-餐厅区占据了房屋中心部分,两端空间服务于更为私密的用途。这项三层设计为使用者提供了不同程度的坚实感。低层的储存室以坚实感为主,上层客厅生活区则几乎完全采用玻璃墙面。上层的卧室则是二者的结合。

Completion Date: Approximately 2018
Area: 975.48sqm
Location: Hutchinson Island, Florida
Landscape Design: Tyler Nielsen

建成时间: 约2018年
面积: 975.48平方米
地点: 佛罗里达州,哈钦森岛
景观设计: 泰勒·尼尔森

Ski House
滑水屋

A group of water skiing enthusiasts had a dream. Banding together with several other fellow water skiers, they purchased agricultural land and modified an existing artificial lake to be used for water skiing/slalom course racing. A house was envisioned for the site.

The designers decided to place the house on the hill dredged from the bottom of the lake. This elevated perspective in an otherwise flat topography provides a view of the watercourse from end to end.

The house is a simple bar, conceptually and literally, broken up to define three areas. The covered platforms serve as a link between pavilions with living space in the middle and bedrooms on either side. A metal roof hovers above these three parts to encourage circulation from one space to another. The roof design also provides shade to the lakeside terrace adjacent to the lap pool. The shape of the roof also captures and directs the water-cooled breezes through the upper transom windows of the living space. This natural air conditioning provides a comfortable setting for most of the year. Rain collection from the valley of the roof is directed to an underground reservoir used for irrigating the native landscape. In the breezeways between the wings, exterior fireplaces on one side and an exterior shower off the master bedroom take full advantage of the Florida climate.

The simplicity of the program, the clarity of the siting, and the design, combined with the elementary materials and means of construction, come together to make a home that is both dramatic and at ease.

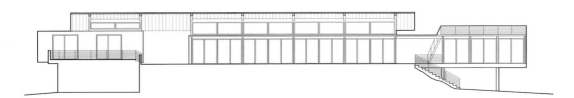

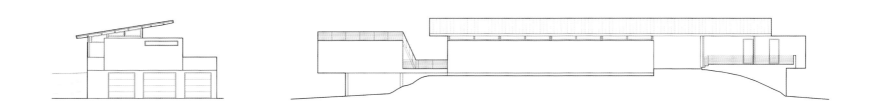

157

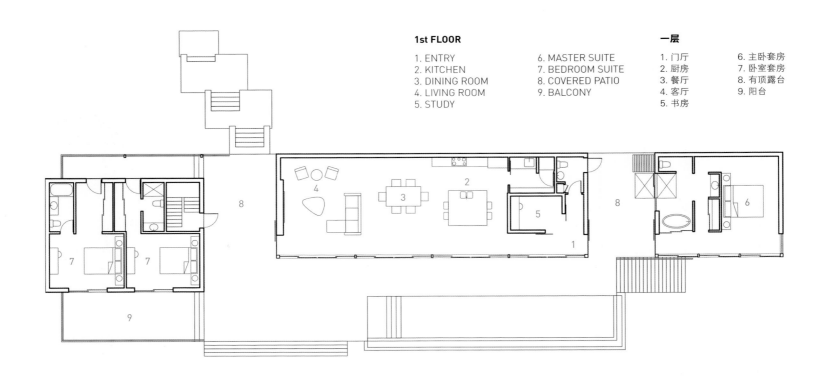

1st FLOOR
1. ENTRY
2. KITCHEN
3. DINING ROOM
4. LIVING ROOM
5. STUDY
6. MASTER SUITE
7. BEDROOM SUITE
8. COVERED PATIO
9. BALCONY

一层
1. 门厅
2. 厨房
3. 餐厅
4. 客厅
5. 书房
6. 主卧套房
7. 卧室套房
8. 有顶露台
9. 阳台

本项目是一群滑水爱好者共同的梦想。几位滑水爱好者一起买下了一块农业用地，将其中一处原有的人工湖改造成滑水及障碍滑水运动场地。还计划在场地上建造一所房屋。

设计师选择将房屋安置在由湖底堆积物打捞上岸形成的小山丘上。相较于原来的平坦地势，抬高后的视角使得水景尽收眼底。

建成后的房屋将是一间简易酒吧，在概念层面与实际意义上划分为三个区域。有棚平台是亭子、中央活动空间与各方向卧室之间的过渡。金属材质屋顶悬吊在这三部分结构之上，方便使用者在不同区域内通行。屋顶设计为使用者提供了通向泳池旁湖滨露台的阴凉通道。屋顶的形状还能将水面上吹来的凉风导向活动区的高处气窗，这种天然空调设计使得室内环境在一年四季都能舒适宜人。屋顶的雨水收集装置导向地下蓄水池，将收集到的雨水用于灌溉当地植被。有棚通道中的室外壁炉和主卧外部的室外淋浴是对佛罗里达州气候的巧妙利用。

项目的简洁设计、选址的明确以及结合了基本材料与基础施工方式的设计共同打造出一个具有夸张个性又让人备感舒适的住宅。

Completion Date: 2004
Area: 334.45sqm
Location: Jupiter Island, Florida
Landscape Design: huum architects
Photography: Tom Winter

建成时间：2004年
面积：334.45平方米
地点：佛罗里达州，朱庇特岛
景观设计：huum建筑事务所
摄影：汤姆·温特

Holly Lane
冬青道

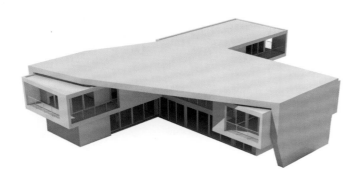

It is unusual for a Florida waterfront property to face south. Our site overlooks a connection to Florida's busy inland waterway. The clients, a young outdoor loving family of four, asked that we find a way to shelter the outdoor space from the mid-day Florida sun.

A dominant, independently supported, "shroud" provides this coverage, as well as creating a frame within which the individual spaces are arranged. Reading like independent structures the rooms appear to hang from the framework. Like a giant ticktacktoe game; some of the boxes are enclosed, some are open.

Additionally, the second floor spaces slide out beyond the first floor footprint to provide covered outdoor spaces around the pool. The all glass living room displays the view of the waterway over the pool whose edge blends with the currents.

Each of the four upper level bedrooms has a private terrace and a framed view of the busy waterway. The master, above and extending beyond the indoor/outdoor family room, appears to hover over the waterway. The upper viewing loggia separates the master from the independent guest suite.

The residence reflects a uniquely Floridian way of living and provides a series of places for gathering and enjoying the serenity of the surroundings, the intimacy of the landscape, and the spirit of the place.

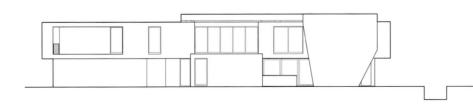
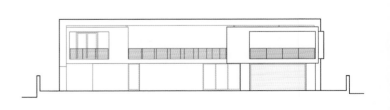

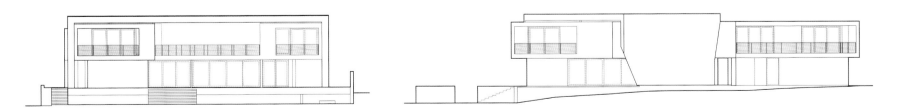

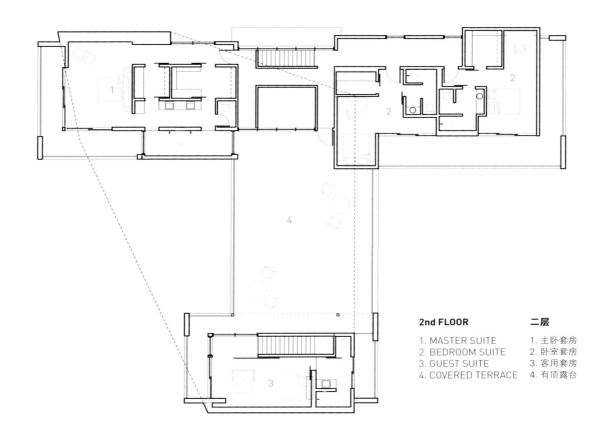

2nd FLOOR 　　　　**二层**
1. MASTER SUITE 　　1. 主卧套房
2. BEDROOM SUITE 　2. 卧室套房
3. GUEST SUITE 　　　3. 客用套房
4. COVERED TERRACE 　4. 有顶露台

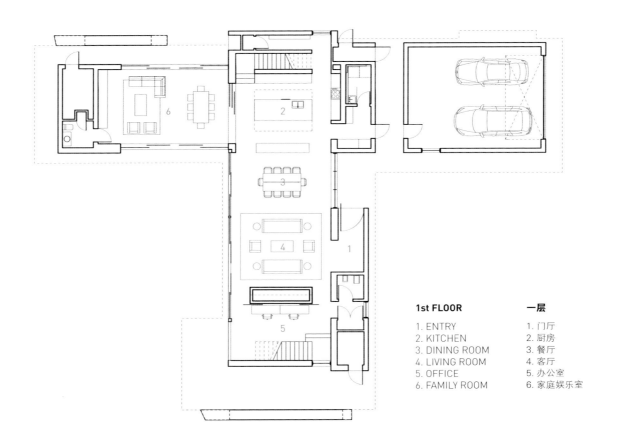

1st FLOOR 　　　　**一层**
1. ENTRY 　　　　1. 门厅
2. KITCHEN 　　　2. 厨房
3. DINING ROOM 　3. 餐厅
4. LIVING ROOM 　4. 客厅
5. OFFICE 　　　　5. 办公室
6. FAMILY ROOM 　6. 家庭娱乐室

在佛罗里达州，朝南的海滨房屋并不多见。本项目中，场地俯瞰佛罗里达州一条繁忙的内陆航道。客户是热爱户外生活的一家四口，希望设计师打造出能够免受佛罗里达州正午烈日炙烤的室外空间。

独立支撑的主体结构提供了这种遮挡，同时形成了独立结构的分布框架。彼此独立的各个房间似乎是悬挂在整体框架上，看起来就像是一个巨大的井字游戏场所；一些空间是封闭的，一些是开放的。

二楼空间在一楼空间的基础上向外平移，在游泳池周围形成有遮挡的室外空间。

客厅采用全玻璃墙面，视野中的泳池与航道水波相交。

上层的四间卧室中每个都有独立阳台，可以欣赏到航道上船只来来往往的情景。位于室内、室外家庭活动空间上方并向外延伸的主卧室似乎悬在航道上方。凉廊将主卧室与客房分隔开来。

这个住宅项目反映了佛罗里达式独特的生活方式，为住户提供了一系列适合社交聚会、欣赏宁静环境与感受空间精髓的场地。

Completion Date: 2016
Area: 641.03sqm
Location: Jupiter, Florida
Landscape Design: huum Architects
Photography: Robin Hill

建成时间：2016年
面积：641.03平方米
地点：佛罗里达州，朱庇特岛
景观设计：huum建筑事务所
摄影：罗宾·希尔

Strulowitz
Strulowitz住宅

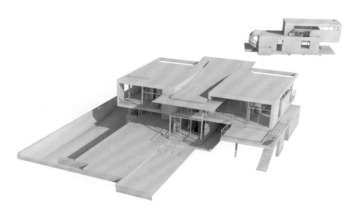

Surrounded by much larger properties this 3,000 sqm site is divided in half by the infrequently traveled local road. On the eastern site, hard against the Atlantic Ocean, is the opportunity to build only on the footprint of the previous structure. On the western side of the street there is the opportunity to build a 464 sqm house. The landscape material required to create visual buffering from adjacent properties greatly limits the buildable area.

The western house is organized as an "upside-down" house with living, dining and master bedroom on the upper level. The rooms open to the east and west, affording dramatic framed morning views of the ocean meeting the horizon, and in the evening, the sun setting over the dense foliage. The entry level has 3 bedrooms and an entry hall gallery to display the owner's art collection. A sloping driveway leads to a 5 bay garage anchoring the basement level. A stainless steel roof plane hovers over the two-story entry hall. This thin blade extends from high above the front door, sloping down into the interior of the main level to separate the public wing from the private, and finally ends as the overhang protecting the main stair from the afternoon sun.

The beach house is an arrangement of stacked volumes sliding past each other to create interior and exterior gathering spaces - inviting engagement while protecting the occupant from the Florida sun. As with the main house, different materials are used top and bottom to promote a horizontal composition complimentary to the always-present horizon line.

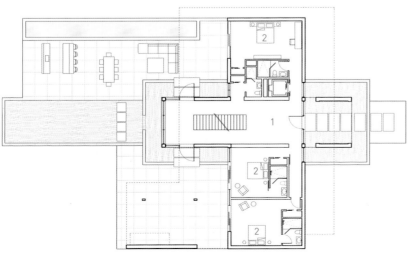

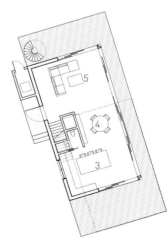

1st FLOOR

1. ENTRY
2. BEDROOM SUITE
3. KITCHEN
4. BREAKFAST NOOK
5. LIVING ROOM

一层

1. 门厅
2. 卧室套房
3. 厨房
4. 早餐角
5. 客厅

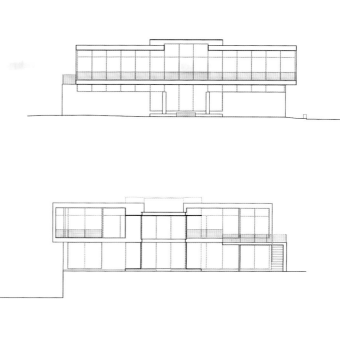

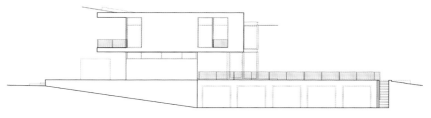

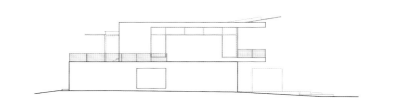

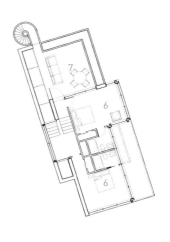

2nd FLOOR

1. ENTRY
2. KITCHEN
3. BREAKFAST NOOK
4. FAMILY ROOM
5. MASTER SUITE
6. BEDROOM SUITE
7. BALCONY

二层

1. 门厅
2. 厨房
3. 早餐角
4. 家庭活动室
5. 主卧套房
6. 卧室套房
7. 阳台

一条很少使用的道路将这块 3,000 平方米的场地一分为二，相比之下四周建筑的面积与规模要大得多。由于场地东侧紧邻大西洋，设计团队只能沿着从前建筑的痕迹建造房屋。街道西侧则可以建造 464 平方米的房屋。针对周围建筑打造视觉缓冲所需的景观材料对可建区域造成了较大限制。

设计师将西侧的房屋打造为"上下颠倒"的设计，即客厅、餐厅和主卧位于楼上。房间朝向东西两侧，住户可以透过东侧窗口体验到大海与天际相接的惊艳景色，傍晚则是落日在繁密的枝叶间下沉的美景。一楼有三间卧室和一个入户走廊，展示主人的艺术品收藏。倾斜的车道通向地下的 5 车位车库。

两层楼高的入户走廊上方是不锈钢屋顶。这层结构从正门上方的高处开始，缓缓地向主要楼层的室内空间倾斜延伸，将公告区域与私密区域分隔，最终在主楼梯处结束，遮挡午后阳光。

这间海滨别墅是多层体量彼此平移的排列设计，借此打造出室内外的聚会空间——鼓励住户社交活动的同时也保护住户免受佛罗里达州烈日侵扰。主要结构中使用不同材料，强调水平构成，与无处不在的地平线形成互补。

Completion Date: Approximately 2019
Area: 715.35sqm
Location: Jupiter Island, Florida
Landscape Design: Innocenti and Webel

建成时间：约2019年
面积：715.35平方米
地点：佛罗里达州，朱庇特岛
景观设计：Innocenti and Webel设计公司

Nests
筑巢

The Summit Powder Mountain development will consist of more than 500 ski accessible home sites within 34 km² of ski slopes. They are connected by a core village with higher density residential buildings, hotels, resident and visitor services, as well as cultural amenities. hughesumbanhowar Architects has designed the prototype for one group of residences equiring a small footprint, densely arranged onto a hillside adjacent to the village center. The neighborhood is the filter between built environment to the west and natural environment to the east. Acknowledging this, the orientation of each unit, in combination with wing walls creates uninterrupted downslope views for each nest, while minimizing openings on the village core side. The overall layout organizes units around the site perimeter, an open central area preserved for a future community amenity. The formal design takes cues from: quartz crystal geometry, local tree species and mountain utility structures. These characteristics provide customization while retaining advantages of a duplicable design.There is a palette of several exterior and interior surfaces that will result in an overall family of related buildings. The Village Nests are designed to meet passive building design standards. Passive building comprises a set of design principles used to attain a quantifiable and rigorous level of energy efficiency within a specific quantifiable comfort level. It makes sense environmentally and financially for these units to minimize energy output to maintain temperature by making the building's envelope more airtight and better insulated rather than spending time and money improving the efficiency of active systems which may amount to looking at a more efficient way of heating outdoor air. The overall design treads lightly on the physical site and operationally minimizing global environmental impact by its energy and resource efficiencies.

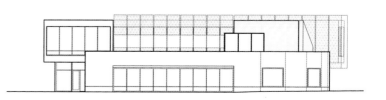

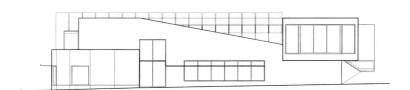
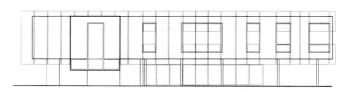

Completion Date: Approximately 2018
Area: 929.03sqm
Location: Hobe Sound, Florida
Landscape Design: Lucido Associates

建成时间：约2018年
面积：929.03平方米
地点：佛罗里达州，霍布桑德
景观设计：卢奇多联合事务所

项目位于南佛罗里达州的铁路轨道和老迪克西高速公路沿线，它所在的小镇距离大西洋仅0.8千米的路程。拆除了有75年历史，已无法修复的老建筑之后人们计划在这块土地上，建造一处与原有结构在大小规模上相近的现代风格建筑。新建筑将综合零售、商业功能。设计团队完成了两个设计方案：第一个方案是在原建筑的基础上进行改进，第二个方案是基于项目及环境的全新设计。从概念层面分析，第一个方案由两个造型独特的结构组成：简单朴素的"种子荚"主体结构以及较小的扭曲造型的"芽"——象征公司本身。从街道角度观察，

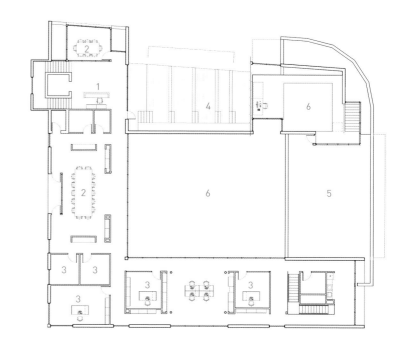

2nd FLOOR   二层
1. RECEPTION   1. 接待处
2. CONFERENCE ROOM   2. 会议室
3. OFFICE   3. 办公室
4. ROOF GARDEN   4. 屋顶花园
5. ROOF DECK   5. 屋顶楼面
6. OPEN TO BELOW   6. 天井

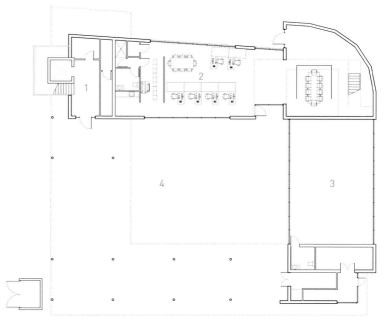

1st FLOOR   一层
1. LOBBY   1. 大厅
2. OFFICE   2. 办公室
3. RETAIL   3. 零售空间
4. GARDEN COURTYARD   4. 花园庭院

主体结构与原建筑较为相似，以此对被拆除的原建筑进行纪念。与原建筑相比，屋顶花园的新设计是唯一的变化。新增的富有表现力的芽形结构被本地植物完全覆盖，植物在阳光下自由生长。这一设计饶有趣味地将整个三楼结构向一侧翻转，插向地面和两层的主体结构。在第二个方案的构想中，两个"L"形楼板反向相对，形成正方形的中央庭院。较高楼层是办公区，将偶尔经过的火车声隔离。较低楼层是零售空间，与街道上的商店相呼应。

Abbott Kinney
阿博特·金尼阁楼

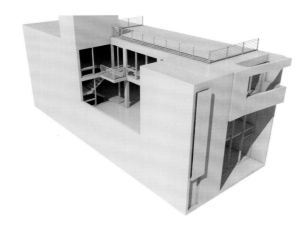

This project is a sustainably conceived and built project on one of Venice Beach's most unique commercial streets.

The building is a 455 sqm, 3 bedroom residence and studio/gallery. The client requested the team layout the building with the first floor serving as a studio/gallery and the remaining floors the family residence.

This storefront window system is continued up the front facade to the second floor, purposely blurring lines of commerce and domesticity. The second floor is the family social area, drawing a thin line of privacy and domesticity from the first floor below - the only method of being closed to the public below is by signage, owing to the clients expectation of frequently entertaining clientele and friends on both first and second floors. Bedrooms are housed on the third floor, the fourth floor roof deck is a private getaway for the family. The neighborhood is densely built and occupied, the client wanted views/sunlight and privacy despite this. By facing south, the courtyard will still receive copious light if and when the adjacent property is built to a similar height and density. By client and architect strategy, this residence implements environmentally sustainable and healthy components and building methods that are intended to be not only conscious of their impact today, but more importantly of the health of the planet and ecosystems of tomorrow.

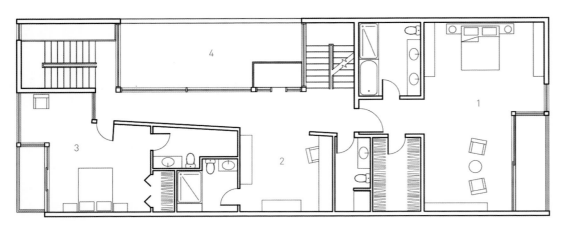

3rd FLOOR   三层
1. MASTER SUITE   1. 主卧套房
2. OFFICE   2. 办公室
3. BEDROOM SUITE   3. 卧室套房
4. OPEN TO BELOW   4. 天井

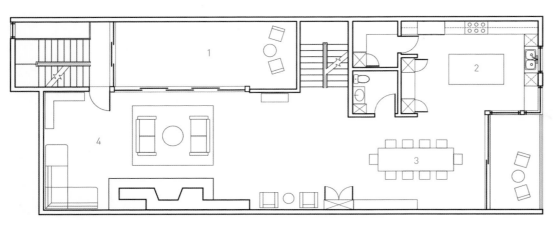

2nd FLOOR   二层
1. COURTYARD   1. 院子
2. KITCHEN   2. 厨房
3. DINING ROOM   3. 餐厅
4. LIVING ROOM   4. 客厅

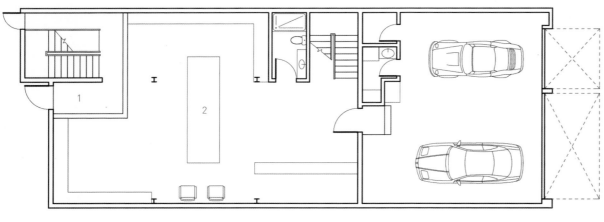

1st FLOOR   一层
1. ENTRY   1. 门厅
2. RETAIL   2. 零售空间

本项目是威尼斯最具特色的商业街上一间拥有可持续设计的建成项目。

项目面积 455 平方米，有 3 间卧室，兼具住宅与工作室或画廊。委托人要求设计团队将一楼打造成工作室/画廊，楼上的其余楼层则用作家庭生活空间。

临街窗口设计沿前立面向上延伸至二楼，有意识地模糊商业与居家之间的界限。二楼的家庭生活空间与下方的一楼空间保留着隐私与家庭生活的微妙界限——由于委托方希望能够经常在一楼和二楼招待客户与朋友，想要与下方的公共空间隔离，唯一的方式就是摆放标志。卧室位于三楼，四楼的屋顶露台是家人进行休闲活动的私人空间。项目所在区域建筑密集，委托人希望能够不受影响，获得良好的视野和隐私性。即便周边建筑到达相近高度和密度，朝南的院子设计仍将接受到充足光照。通过委托方和建筑师的策略，这间住宅使用了可持续且健康的组成部件，这样的建筑方式不仅关注项目对当下的影响，更重要的是关注了地球的健康与生态系统的未来健康。

Completion Date: 2007
Area: 455.22sqm
Location: Venice, California
Landscape Design: huum architects
Photography: Nick Springett

建成时间：2007年
面积：455.22平方米
地点：加利福尼亚州，威尼斯
景观设计：huum建筑事务所
摄影：尼克·斯伯格蒂

Prairie Chapel
草原小教堂

The clients requested that the designers reuse corrugated metal panels and wood structural elements from a recently razed quonset hut. They wanted a destination on 0.16km² of former grazing land now being restored to native prairie. It will be used for memorials, celebrations and other ceremonies. The basic arched form of the quonset hut is a pinched, twisted, and sliced room. Like a standard farm shed, the tactic for this structure was to insert one type of altered farm building inside another. The wood charring is a Japanese technique designed to provide a fire resistant outer layer to the wood, a preburned effect that tempers flammability. The prairie is maintained by burning it every few years to allow new seeds to sprout and clear away dead vegetation. In addition to this charred wood, the building is covered and then reassembled. The original full-length arched trusses are modified and re-assembled with raw metal plates to produce the new composite curved frames. A continuous 0.6m wide gap in the ceiling becomes the entry into the south facing wall, leaving the interior partially exposed to the elements. When a visitor enters this gap, they abruptly encounter the wall of another room, a porous slatted envelope made from charred and reused cedar siding. One open corrugated metal wall surrounded by a 0.9m gravel limestone walk renders the building fireproof. This same local limestone serves as the interior flooring of the building. Snowstorms and full moons will make for very magical nights within the space.

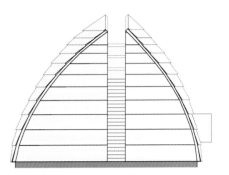
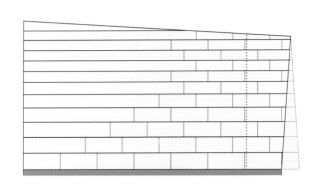

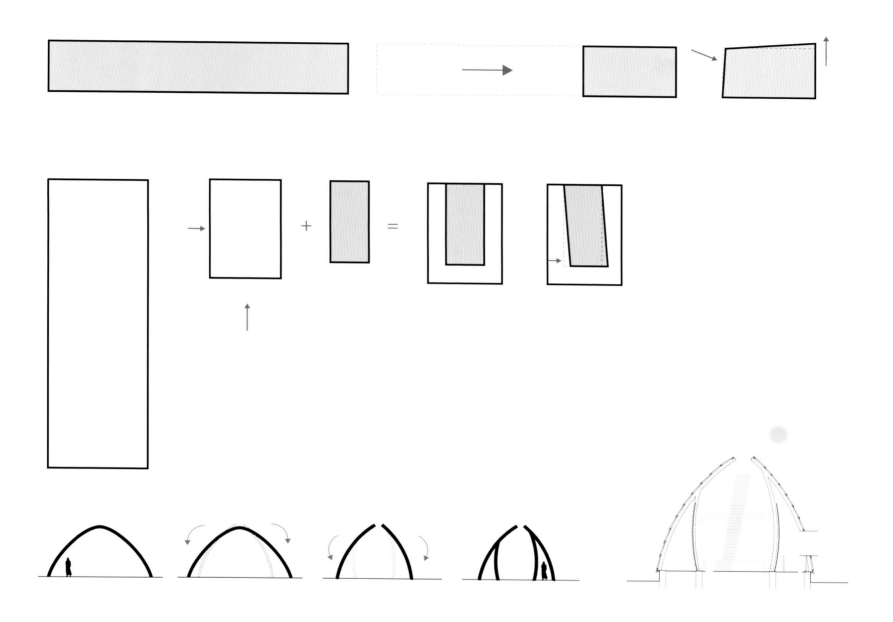
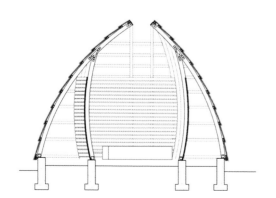
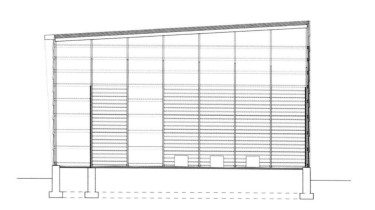

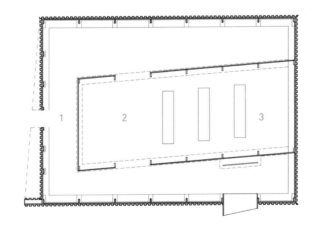

1st FLOOR

1. ENTRY
2. SKY SANCTUM
3. PEWS

一层

1. 门厅
2. 圣所
3. 长凳

客户要求设计团队在这个项目中重新利用近期从一间圆拱屋上拆除的波纹金属板和木质材料。项目所在的这块 0.16 平方千米场地曾经作为牧场使用，如今被恢复成草原。客户希望在这里建造一个用于举行纪念仪式、庆祝活动以及其他仪式的场所。圆拱型活动房屋的基本弓形是通过修剪、扭转、切割和组装实现的。与标准农场小屋中的开放空间不同，本项目中的板条墙壁结构似乎是将一种改良后的农场结构放置在另一个建筑中。木材炭化是一项用于为木材增添防火外层的日本技术，产生的预燃效应可以改变材料的阻燃性。为了维护草原的生态健康，每隔几年就会进行一次烧荒，促进新种子发芽，同时清除已经死去的植被。除了炭化木材，项目还使用了波纹金属覆面。设计师对原来的全长拱形桁架进行调整后利用粗金属板材重新组装，打造全新的复合式曲线框架。通过天花板上连续的 0.6 米宽缝隙，就会来到朝南的墙壁前。访客进入这个空间时，会突然面对另一个房间的多孔板条外墙，材料为焦化的回收雪松壁板。在周围铺设了 0.9 米宽的砂砾岩小路，提高了建筑在烧荒中的防火性能。设计师还选择了相同的当地石灰石作为室内地面材料。在这里度过暴风雪夜和满月的夜晚将会感觉非常奇妙。

Completion Date: 2012
Area: 53.88sqm
Location: Minnesota
Landscape Design: huum architects

建成时间：2012年
面积：53.88平方米
地点：明尼苏达州
景观设计：huum建筑事务所

Augusta
奥古斯塔

A family's dream of a special place for present and future generations has been realized on a Florida barrier island. While the shoreline cannot be seen from the property, the landscape of the 8,000 sqm site offers beautiful views in every direction. A winter retreat for a couple and their adult children, the focus on detail and order is a reflection of their European heritage and modern lifestyle. Conceived as a group of buildings built and connected over time, Augusta is a flowing composition of open and light-filled spaces. The oversized roof, supported by stylized structural concrete columns, shelters two parallel universes: inside spaces adjacent to similar-sized covered outdoor spaces. The interior living room leads to the outdoor living room; this pattern repeats the length of the main house. While only the master bedroom is located in the main house, two identical guesthouses, each equally separate from the other, turn 90 degrees to face the oversized swimming pool centered on the width of the main house. Carefully considered details compliment every interior surface. No opportunity to celebrate the meeting of different materials is missed. Clear cedar ceiling planes contrast with the white plaster, emphasizing sharp 90-degree edges and defining the circle of the dramatic stairwell. Smooth plaster is a prominent player in this composition as it twists up the main staircase, floats above the floors on structural columns or stops short of meeting the ceilings at the continuous line of transom windows.

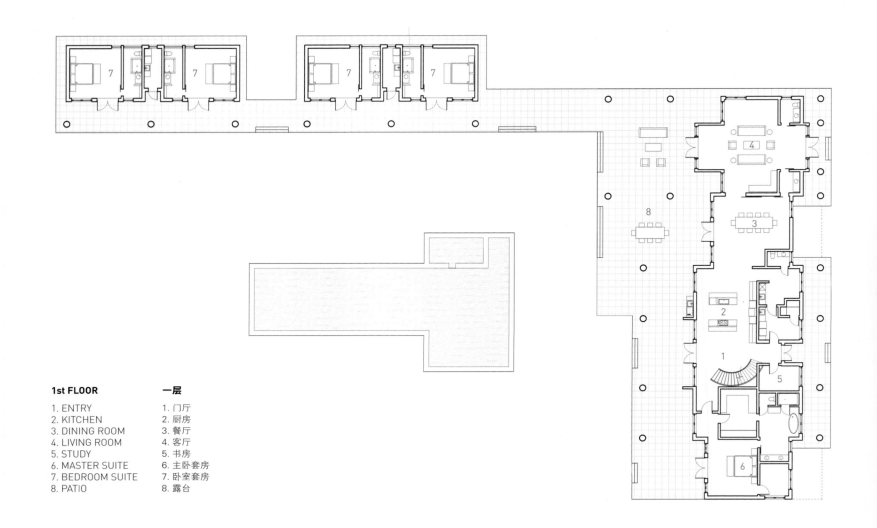

1st FLOOR   一层

1. ENTRY   1. 门厅
2. KITCHEN   2. 厨房
3. DINING ROOM   3. 餐厅
4. LIVING ROOM   4. 客厅
5. STUDY   5. 书房
6. MASTER SUITE   6. 主卧套房
7. BEDROOM SUITE   7. 卧室套房
8. PATIO   8. 露台

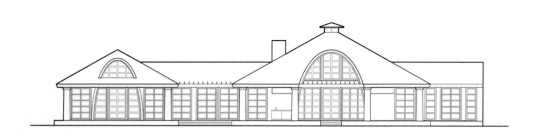

佛罗里达堰洲岛上的这个项目凝聚了一个家庭对于现在和未来几代人拥有一个特别住所的梦想。虽然在住宅中无法看到海岸线，占地 8,000 平方米的场地本身就提供了全方位的秀丽风景。这里是一对夫妇和他们的成年子女冬季度假的场所，对细节和秩序的关注反映了他们的欧洲传统和现代生活方式。奥古斯塔作为一组随着时间推移建成并联系在一起的建筑，呈现出充分开放且光线充足空间的流线组合。尺寸夸张的屋顶由风格化的混凝土立柱支撑，下方是两个平行空间：室内空间及与其相邻且尺寸相当的有顶室外空间。室内客厅通向室外客厅，这种模式沿住宅主体的长度重复。项目中仅主卧室位于住宅主体结构中，两个完全相同的客房彼此分开，以 90 度角朝向与住宅主体中心对齐的超大游泳池。精心设计的细节使得每个室内布局都耐人寻味。设计师没有错过任何一个让不同材料交相辉映的机会。杉木天花板与白色石膏形成反差，突出 90 度的尖锐边缘和造型夸张的楼梯井。光滑的石膏造型随主楼梯旋转而上，悬在楼层之间，在空间构成中发挥了重要作用。

Completion Date: 2008
Area: 492.39sqm
Location: Jupiter Island, Florida
Landscape Design: Innocenti and Webel
Photography: Allan Carlisle Photography

建成时间：2008年
面积：492.39平方米
地点：佛罗里达州，朱庇特岛
景观设计：Innocenti and Webel设计公司
摄影：艾伦·卡莱尔摄影公司

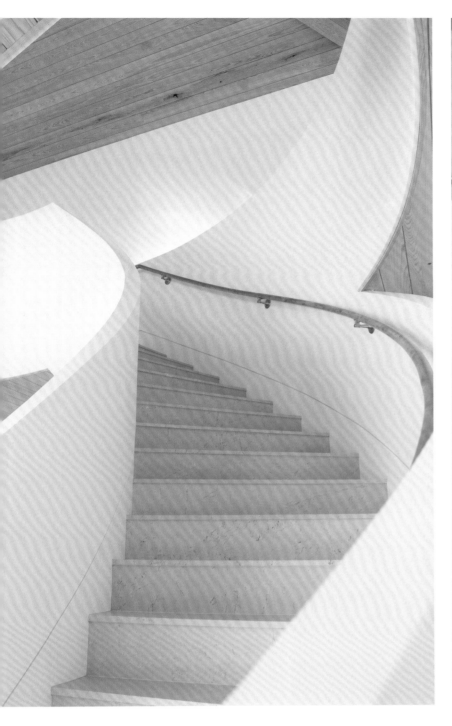

Erik
埃里克

At the edge of a nature preserve, "Erik" is an oceanfront home on a wide sandy beach with sweeping views of the eastern horizon.

Built for a husband and wife planning to entertain frequently, the residence is divided into three two-story pavilions connected by one - story hallways that in plan form a leeward side pool courtyard. A broken roof line diminishes the building's mass, while a louvered overhang extends the length of the house, shielding the living spaces from the arc of the sun.

Cantilevered slabs of the second level are connected to the limestone pavers at ground level by expanses of floor to ceiling glass, floating the upper volumes and further lightening the building's presence. 2 story high isolated stone walls anchor building to ground at defining locations.

Pavilion adjacencies are organized by the amount of time they're occupied--the least used southern entry pavilion has two upper guest rooms and an office and media room below; a single 2-story living room occupies the middle pavilion; the north pavilion is where day to day living occurs- garage, kitchen, family room and the master suite stacked one above the other.

Post - tensioned concrete beams allow an unobstructed view of the ocean from all of the eastern rooms. The western pool fills a courtyard created by the pool cabana and sun sheltering breezeway. The vast living room with its undulating ceiling provides numerous views and seating areas, some intimate, some expansive.

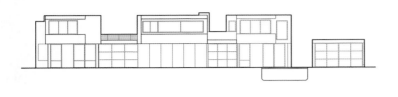
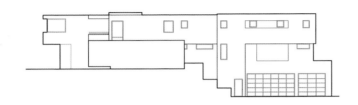
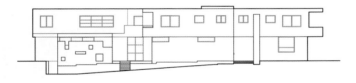
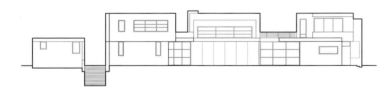
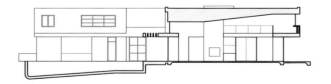

项目位于自然保护区边缘，是坐落在宽阔沙滩上的海滨住宅，乐享东方地平线的宜人景色。

住宅的主人是一对夫妇，他们计划经常来到这里休闲放松。设计师将住宅分成三个两层结构，由一个单层走廊连通，形成背风并配备了游泳池的院子。中断的屋檐线使得住宅看起来比实际稍小，百叶吊帘对房屋的长度加以延伸，遮挡直接射入客厅的日光。

落地玻璃窗一端是二楼的悬臂式平板，一端是一楼的石灰石铺面，赋予上层结构飘浮感的同时也进一步提亮整个空间。两层高的独立石墙在关键位置提供支撑。

结构单元之间按照使用时间进行组织排序：最少使用的南侧入户大厅上层有两间客房和一间办公室，下层有一间多媒体室；中间单元是独立的两层客厅；北侧是使用频率最高的日常生活空间——车库、厨房、家庭活动室和主卧，依次向上堆叠。

后张混凝土梁的应用使得朝东的房间可以饱览开阔的海景。住宅西侧的游泳池以及池边小屋使院内空间变得充实。宽敞的客厅配合起伏天花板为住户提供了或私密或开阔的变化视野和不同的座位空间。

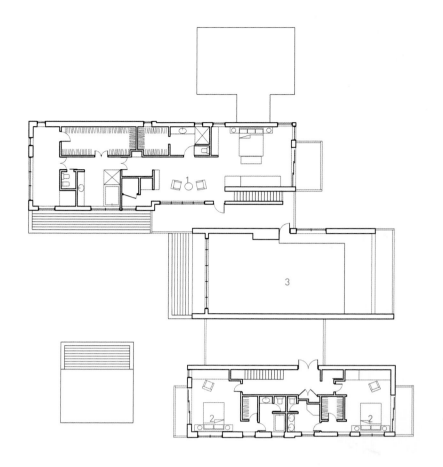

2nd FLOOR　　　二层

1. MASTER SUITE　　1. 主卧套房
2. BEDROOM SUITE　 2. 卧室套房
3. OPEN TO BELOW　 3. 天井

1st FLOOR　　　一层

1. ENTRY　　　　1. 门厅
2. KITCHEN　　　2. 厨房
3. DINING ROOM　3. 餐厅
4. FAMILY ROOM　4. 家庭活动室
5. LIVING ROOM　5. 客厅
6. OFFICE　　　　6. 办公室
7. MEDIA ROOM　 7. 多媒体室
8. GUEST SUITE　 8. 客用套房

Completion Date: 2002
Area: 929.03sqm
Location: Jupiter Island, Florida
Landscape Design: huum architects
Photography: Ken Hayden

建成时间:2002年
面积:929.03平方米
地点:佛罗里达州,朱庇特岛
景观设计:huum建筑事务所
摄影:肯·海登

图书在版编目（CIP）数据

水之匠心：休斯乌姆班霍瓦尔建筑事务所作品集／
（美）斯科特·休斯，（美）约翰·乌姆班霍瓦尔著；
张晨译. — 沈阳：辽宁科学技术出版社，2017.3
　ISBN 978-7-5591-0014-6

Ⅰ.①水… Ⅱ.①斯… ②约… ③张… Ⅲ.①建筑
设计－作品集－美国－现代 Ⅳ.① TU206

中国版本图书馆CIP数据核字(2017)第 300464 号

出版发行：辽宁科学技术出版社
　　　　　（地址：沈阳市和平区十一纬路25号 邮编：110003）
印 刷 者：上海利丰雅高印刷有限公司
经 销 者：各地新华书店
幅面尺寸：254mm×305mm
印　　张：21
插　　页：4
字　　数：200千字
出版时间：2017年3月第1版
印刷时间：2017年3月第1次印刷
责任编辑：杜丙旭　于峰飞
封面设计：周　洁
版式设计：周　洁
责任校对：周　文

书　　号：ISBN 978-7-5591-0014-6
定　　价：268.00元

联系电话：024-23280367
邮购热线：024-23284502
http://www.lnkj.com.cn